LOGIC AND BOOLEAN ALGEBRA

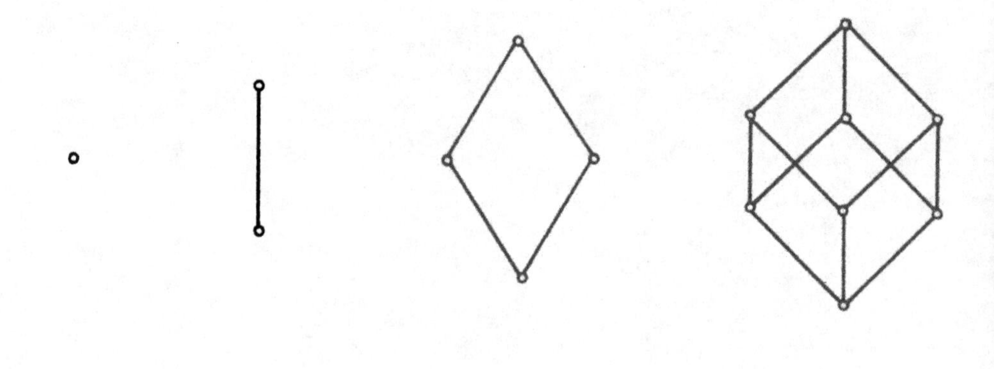

LOGIC AND BOOLEAN ALGEBRA

B. H. Arnold
University of Oregon

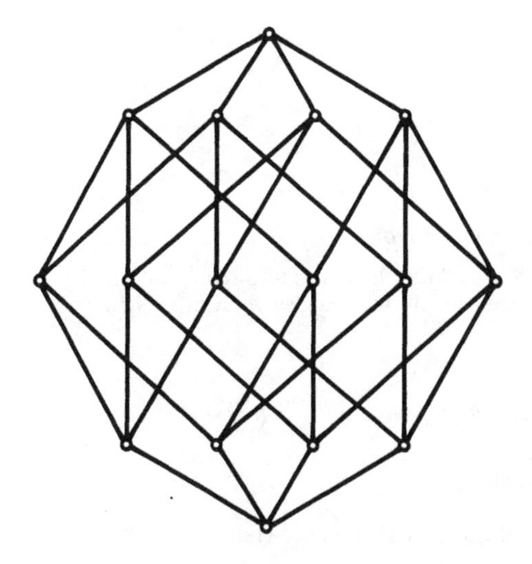

Dover Publications, Inc., Mineola, New York

Bibliographical Note

This Dover edition, first published in 2011, is an unabridged republication of the work originally published by Prentice-Hall, Inc., Englewood Cliffs, N.J., in 1962.

Library of Congress Cataloging-in-Publication Data

Arnold, B. H. (Bradford Henry), 1916–
Logic and boolean algebra / B. H. Arnold.
p. cm.
Originally published: Englewood Cliffs, N.J. : Prentice-Hall, 1962.
ISBN-13: 978-0-486-48385-6
ISBN-10: 0-486-48385-1
1. Algebra, Boolean. I. Title.

QA10.3.A76 2011
511.3'24—dc23

2011017585

www.doverpublications.com

PREFACE

This text has evolved from successive revisions of notes used in a one-quarter course at the upper division level. Additional material has been included so that the text may be used for a semester course.

Chapter 1 presents the rudiments of intuitive logic, but this subject matter serves mainly to provide illustrations of the algebraic structures discussed later. Chapter 2 introduces Boolean functions (functions whose variables have the set $\{0, 1\}$ as a range), discusses equivalence relations, and shows that the Boolean functions of n independent variables can be used to define an equivalence relation in the set of truth value functions of n independent variables.

The main purpose of the book is to develop the theory of Boolean algebras via the study of simpler algebraic systems. Chapter 3 begins this study with a very simple case — (partially) ordered sets. Definitions are given for an arbitrary algebraic system and for an isomorphism between two such systems. Finite ordered sets are represented graphically. In Chapter 4 the ordered sets are specialized to lattices. The distributive law and complementation are discussed. A lattice is characterized in terms of the binary operations sup and inf.

In Chapter 5 a Boolean algebra is defined as a complemented, distributive lattice and it is shown that every finite Boolean algebra is isomorphic to the algebra of all subsets of some set. Boolean rings are treated in Chapter 6; it is proved that Boolean algebras are equivalent to Boolean rings with a unit element, and that every finite Boolean ring can be represented as the set of all n-tuples of Boolean constants (for some n). These results give another representation theorem for finite Boolean algebras. Chapter 7 discusses the disjunctive and conjunctive normal forms in a Boolean algebra as well as the ring normal form in a Boolean ring. Both existence and uniqueness of these normal forms are studied. The principle of duality is postponed to the last section of Chapter 7 for two reasons. First, duality is the source of several problems stated earlier in the text, and it is hoped that the student will carefully think through the proofs of the corresponding theorems in the text when solving these problems. Secondly, the better students will discover the principle of duality for themselves.

In the first seven chapters, the only applications considered are to set theory or logic. Chapter 8 presents a brief introduction to the uses of Boolean algebra in the design and analysis of switching circuits and computers. Some problems of the logical puzzle type are also solved.

The problems are considered an integral portion of the text. The serious student should do most of them. An occasional harder problem is marked with an asterisk, "*"; problems marked with a "#" give results which are referred to later in the text. The symbol "▮" is used in the text to denote the end of a proof. As a guide to further reading, a short bibliography is included which lists references at all levels from freshman texts to current research papers.

The author is indebted to his students and colleagues for their many helpful suggestions during the incubation period of this text. It would be impossible to list all who have contributed, but it would also be improper to omit mention of the few who have contributed most. Miss Sandra Anderson, Professor Harry E. Goheen, and Miss Patricia Prenter each read the entire manuscript, at various stages. Of course, any remaining errors are the sole responsibility of the author.

B. H. ARNOLD

CONTENTS

LOGIC AND BOOLEAN ALGEBRA

1

SOME CONCEPTS OF INTUITIVE LOGIC

1-1 Introduction

The portion of logic with which we shall be concerned deals with sentences, certain structures based on sentences, and methods of proof. Much of the material presented in this chapter will reappear in later chapters, approached from a different viewpoint. The technical term "sentence" is defined in Section 1-2 and the most common connectives used to combine sentences to produce others are discussed.

The algebraic concepts of variable and function are reviewed in Section 1-3 and are specialized to sentences to define sentential functions. Truth tables are introduced in Section 1-2 and are further developed in Section 1-3.

Section 1-4 discusses relations between sentences and between sentential functions. The important relations of implication and equivalence are contrasted with two of the connectives introduced in Section 1-2. Truth value functions are defined as a special type of sentential function.

Some of the most common methods of proof are given in Section 1-5. Special emphasis is given to proving an implication since many theorems in mathematics are stated in this form. Direct proofs and several types of indirect proofs of implications are discussed. The method of proof by mathematical induction is also discussed; this method is applicable to theorems of a type that occurs rather frequently in mathematics.

1-2 Sentences and Connectives

In the study of English grammar, sentences are divided into several types — declarative, interrogative, exclamatory, etc. The sentences we shall study are of the declarative type; these are the sentences usually used to convey information. In ordinary usage, when such a sentence is written the author conveys that the sentence is true or at least that he believes it to be true. However, the subject for a debate might be stated as a declarative sentence and of course the printing of this sentence in the announcement of the debate should not be taken as an indication that the statement is actually true. In logic, the term "sentence" is used in a technical sense as described in the following definition.

DEFINITION 2.1 A *sentence* is a statement that is either true or false but not both.

Notice that it is not necessary to know whether the statement is true or whether it is false; the only requirement is that it should definitely be one or the other. In some cases the truth or falsity of a sentence may be well known; in others it may be possible to reach a decision after doing some work — perhaps for a few moments or for a few years; in still other cases it may appear impossible to reach a decision. If a decision is reached, the statement under consideration is a sentence, but even in cases where a decision is not reached it may be possible to argue that the statement is a sentence but that it just doesn't happen to be known whether it is a true sentence or a false one. A few examples will clarify this point.

Example 2.2 Each of the following is a sentence.

(a) New York is a city in the United States of America.

(b) $2 \cdot 2 = 5$.

(c) $\sin \pi = -1$.

(d) The digit in the 503rd decimal place in the decimal expansion of π is a 3.

Example 2.3 None of the following is a sentence.

(a) Of at the by from tomorrow.

(b) Come here!

(c) Are you hungry?

Given a supply of sentences a, b, ... there may be a great many ways in which they can be combined to form new sentences. Fortunately, we may limit ourselves to the successive use of relatively few simple combinations. Moreover, some of these simple combinations are quite familiar from everyday usage. We shall use five of these simple combinations although, for many purposes, three of them could be dispensed with, as we shall see in Section 1-4. The 5 connectives used to form these 5 basic combinations are "*not*", "*and*", "*or*", "*if* ... *then* ...", and "*if and only if*". We shall discuss each one in turn, but first it is convenient to introduce a special notation.

The numbers 1 and 0 are called *truth values;* we say that the truth value of a true sentence is 1 and that the truth value of a false sentence is 0. This numerical notation will sometimes be more convenient than the words "true" and "false" or the letters "T" and "F" which are used by some authors. The definition of a sentence can be rephrased, with this new notation, as: A sentence is a statement whose truth value is 1 or 0, but not both.

The connective "*not*" is applied to a single sentence "a" to form the sentence "not a" called the *negation* of "a", in symbols, $\sim a$. Of course, the sentence "$\sim a$" states that the sentence "a" is false; "$\sim a$" may be false, in which case, "a" is true. Since the defining characteristic of a sentence is that it should be either true or false but not both, it is useful to have a table which shows the truth value of "$\sim a$" for each possible truth value of "a." Figure 2.1 shows such a table; it is called a *truth table* for "$\sim a$". Since "a" is given as a sentence, its truth value must be either 1 or 0 and cannot be both. From Fig. 2.1 it can be seen that "$\sim a$" also enjoys this property, so that "$\sim a$" is indeed a sentence.

The connective "*and*" is applied to two sentences "a" and "b" to form the sentence "a and b" called the *conjunction* of "a" and "b", in symbols

FIGURE 2.1

a	$\sim a$
1	0
0	1

a	b	$a \wedge b$
1	1	1
1	0	0
0	1	0
0	0	0

FIGURE 2.2

a	b	$a \vee b$
1	1	1
1	0	1
0	1	1
0	0	0

FIGURE 2.3

$a \wedge b$. The sentence "$a \wedge b$" states that both of the sentences "a" and "b" are true (of course, "$a \wedge b$" may be false). Figure 2.2 shows its truth table. Notice that there are 4 rows in this table since there are 4 possibilities for the pair of truth values of the sentences "a" and "b." The information in Figs. 2.1 and 2.2 is perfectly familiar.

The connective "*or*" is applied to two sentences "a" and "b" to form the sentence "a or b" called the *disjunction* of "a" and "b", in symbols $a \vee b$. Figure 2.3 shows its truth table. In ordinary English usage the word "or" is used in two different senses. Sometimes it means that exactly one of two alternatives occurs and sometimes it means that at least one occurs. In logic, the second of these two meanings has been adopted as standard. The first row of Fig. 2.3 shows that "$a \vee b$" is true in the case where both of the sentences "a" and "b" are true.

The connective "*if . . . then . . .*" is applied to two sentences "a" and "b" to form the sentence "if a then b" called the *conditional* of "a" and "b", in symbols $a \rightarrow b$. Figure 2.4 shows its truth table. Notice that in case "a" has truth value 0 the conditional "$a \rightarrow b$" is automatically true no matter what sentence "b" is used. In case "a" has truth value 1 there is a further requirement in order that the conditional "$a \rightarrow b$" should be true; the first two rows in Fig. 2.4 show that this requirement is just that "b" should be true. These facts usually seem a little surprising at first acquaintance, and it is important that the student should arrange his intuition to agree with Fig. 2.4. One possible way to accomplish this is to think of the requirement for the truth of the conditional "$a \to b$" as follows:

> If "a" is true, we require that "b" should be true;
> if "a" is false, no requirement is made.

Two further remarks should be made regarding the conditional. First, in ordinary conversation, the sentence "if a then b" would frequently indicate that there is a relation of cause and effect between the situations described in the sentences "a" and "b." This is *not* the sense in which the conditional is used in logic. The conditional "$a \to b$" can be formed with any two sentences "a" and "b;" they do not have to be related in any way.

Second, a conditional is a sentence and may therefore be either true or false. The mere stating of a conditional "$a \to b$" does not mean that the sentence is true, and it certainly does not mean that either "a" or "b" individually is true. In this connection, it is important to notice that two different "languages" are being used in this text. Many of our theorems will be phrased in the form "if . . . , then . . . ;" this phrasing will also occur in proofs of theorems, etc. In such places, the statement appears in our *metalanguage*, and we mean to indicate that the sentence is actually true. In other occurrences, a conditional may merely be presented for consideration or study, and no indication will be given as to whether it is true or false; in these places, the statement appears in our *object language*. Similar remarks apply to sentences in forms other than the conditional. The context will usually make clear which meaning is intended, but a special notation will be introduced in Section 1-4 for use in the metalanguage.

Some examples will clarify the distinction between these two languages. In the metalanguage, we may write the sentence "2 + 3 = 5." This sentence gives us some information about the numbers 2, 3, and 5. This same sentence would appear in our object language if we wrote " '2 + 3 = 5' is a sentence about integers." This latter sentence gives us no informa-

FIGURE 2.4

a	b	$a \to b$
1	1	1
1	0	0
0	1	1
0	0	1

a	b	$a \leftrightarrow b$
1	1	1
1	0	0
0	1	0
0	0	1

FIGURE 2.5

tion about the sum of 2 and 3; it tells us something about the sentence "$2 + 3 = 5$." A typical occurrence of a sentence in the object language would be as "The sentence 'a' is true." This gives us the same information as "a" in the metalanguage.

The connective "*if and only if*" is applied to two sentences "a" and "b" to form the sentence "a if and only if b", called the *biconditional* of "a" and "b" in symbols $a \leftrightarrow b$ or a iff b. Figure 2.5 shows its truth table. Remarks similar to those made about the conditional apply to the biconditional. The requirement for the truth of the biconditional "$a \leftrightarrow b$" is as follows:

> Both of "a" and "b" must be true or both must be false, but we don't care which.

The biconditional is a sentence and may be either true or false. The writing of "$a \leftrightarrow b$" does not indicate that the sentence is true and certainly does not mean that either "a" or "b" individually is true. However, statements of the form ". . . iff . . ." will appear in our metalanguage (especially in definitions) and, in these occurences, we do mean to indicate that the statement is true. The context will usually make clear which meaning is intended, but a special notation will be introduced in Section 1-4 for use in the metalanguage. The biconditional can be formed with any two sentences; no relation of cause and effect is indicated.

Example 2.4 Let "a" be the sentence "New York is a city in the United States of America", let "b" be the sentence "$2 \cdot 2 = 5$", and let "c" be the sentence "$\sin \pi = -1$".

(a) "$\sim a$" has truth value 0.

(b) "$\sim b$" has truth value 1.

(c) "$a \wedge b$" has truth value 0.

(d) "$a \wedge a$" has truth value 1.

(e) "$a \vee b$" has truth value 1.

(f) "$a \rightarrow b$" is a false sentence.

(g) "$c \rightarrow a$" is a true sentence.

(h) "$b \leftrightarrow c$" is a true sentence.

(i) "$a \rightarrow [b \vee \sim (c \wedge a)]$" is a true sentence.

Example 2.4(i) illustrates the use of parentheses of various types to indicate the order in which the connectives operate when several are involved in the same sentence. When the order of operation is immaterial, parentheses may be omitted. To further decrease the need for parentheses we follow the usual convention that the connective "$\sim$" binds more strongly than either "$\wedge$" or "$\vee$", and each of these in turn binds more strongly than either "$\rightarrow$" or "$\leftrightarrow$". Example 2.5 illustrates this convention.

Example 2.5 Let "a", "b", and "c" be sentences.

(a) "$(\sim a) \wedge b$" may be written "$\sim a \wedge b$."

(b) "$[(\sim a) \vee b] \rightarrow c$" may be written "$\sim a \vee b \rightarrow c$."

(c) "$(a \wedge b) \vee c$" requires parentheses.

(d) "$a \vee (b \vee c)$" may be written as "$a \vee b \vee c$."

(e) "$a \rightarrow (b \leftrightarrow c)$" requires parentheses.

PROBLEMS

1. Which of the following are sentences? In each case try to tell why you think that the expression is or is not a sentence. What are the respective truth values?

   (a) I am beautiful.

   (b) You are beautiful.

   (c) Let there be light.

   (d) The moon is made of green cheese.

   (e) There is intelligent life on Mars.

*(f) This sentence is false.

*(g) This sentence is true.

(h) Mathematics is easy.

(i) $\frac{1}{2} + \frac{1}{4} = \frac{3}{4}$.

(j) $\frac{1}{2} + \frac{1}{4} + \frac{1}{8} + \ldots = 1$.

(k) How is Mary?

(l) The digit 3 appears more than 50 times in the decimal expansion of π.

(m) The digit 3 appears an infinite number of times in the decimal expansion of π.

2. Let "a" be the sentence "$3 > 5$", let "b" be the sentence "$3 + 2 = 5$", and let "c" be the sentence "$3 + 5 = 2$". What is the truth value of each of the following sentences?

(a) $a \wedge \sim b$.

(b) $\sim (a \wedge b)$.

(c) $\sim a \wedge \sim b$.

(d) $\sim (a \vee b)$.

(e) $a \wedge b$.

(f) $a \rightarrow c$.

(g) $b \leftrightarrow c$.

(h) $a \rightarrow (b \rightarrow c)$.

(i) $(a \rightarrow b) \rightarrow c$.

(j) $c \rightarrow \sim\{[a \wedge (b \rightarrow a)] \vee [(c \vee b) \rightarrow a]\}$.

(k) $(a \vee b \vee c) \leftrightarrow (a \wedge b \wedge c)$.

3. Suppose there are a particular square and a particular circle under consideration. Let "a" be the sentence "The square is red or the circle is larger than the square." Let "b" be the sentence "If the square is smaller than the circle then the circle is blue." Let "c" be the sentence "The square is blue if and only if the square and circle are different colors." For each of the following sentences find, if possible, several examples of squares and circles for which the sentence is true and several examples for which it is false.

(a) $a \wedge b$.
(b) $a \wedge c$.
(c) $b \wedge c$.
(d) $a \rightarrow b$.
(e) $b \rightarrow a$.
(f) $a \leftrightarrow c$.
(g) $b \leftrightarrow c$.
(h) $a \vee b \vee c$.
(i) $a \rightarrow (a \leftrightarrow c)$.
(j) $(a \rightarrow a) \leftrightarrow c$.
(k) $(a \wedge b) \vee c$.
(l) $a \wedge (b \vee c)$.

4. Let "a" be the sentence "George is richer than Harry", let "b" be the sentence "George is taller than Harry", and let "c" be the sentence "Harry is taller than George." For each of the following sentences, what information are you given about Harry and George if you are told that the sentence has truth value 1? Express the information in a convenient form.

(a) $a \vee b$.
(b) $a \wedge b$.
(c) $\sim a \vee b$.
(d) $\sim (a \vee b)$.
(e) $b \wedge c$.
(f) $\sim b \wedge \sim c$.
(g) $a \wedge (a \rightarrow b)$.
(h) $\sim a \wedge (a \rightarrow b)$.
(i) $b \rightarrow c$.
(j) $a \leftrightarrow (b \vee c)$.
(k) $b \wedge (a \rightarrow b)$.
(l) $\sim b \wedge (a \rightarrow b)$.
(m) $a \wedge b \wedge c$.
(n) $a \vee b \vee c$.
(o) $(a \vee b) \wedge c$.
(p) $a \vee (b \wedge c)$.

1-3 Variables and Functions

In this section we review some of the concepts of elementary algebra and formulate the definitions in a way that will be convenient for use in our study of logic.

DEFINITION 3.1 A *variable* is a symbol together with a set of objects, and it is understood that the symbol stands for any element of the set. The set is the *range* of the variable and each element of the set is a *value* of the variable.

In describing a particular variable it frequently happens that only the symbol is mentioned, and it is left to the reader to gather, from context or experience, what set is being used for the range of the variable. For example, if the expression "$1/x$" is under consideration one might speak of the variable x, or even of the real variable x, and expect the reader to understand that the range of the variable is the set of all non-zero real numbers. Such inexact descriptions of variables do not seem to result in much confusion, but whenever a new variable is introduced it is worth-while to give some consideration to what set is being used as its range.

Example 3.2 Each of the following is a variable.

(a) The symbol is n; the range is the set of all positive integers.

(b) The symbol is ⊓ ; the range is the set of all chairs in the Radio City Music Hall in New York.

(c) The symbol is 4; the range is $\{4\}$. A variable whose range contains only one element is called a *constant.* The constants are a particularly important type of variable.

In our further work with sentences we shall need to discuss any one of a set of sentences; the concept of a variable enables us to do this conveniently. A *sentential variable* is a variable whose range is a set of sentences. To avoid exceptional cases in some of our later theorems, we require that the range of each sentential variable contain at least one true sentence and at least one false sentence; the range will usually be the set of all sentences.

A single variable can be studied at considerable length using statistical methods but we shall be more interested in studying relationships among several variables. A particularly important relationship is given by the following definition.

DEFINITION 3.3 A variable z is *functionally related* to the n variables $x_1, x_2, \ldots, x_n$ iff there is a procedure which determines a unique value of z whenever a value for each of $x_1, x_2, \ldots, x_n$ is given. The variable z is the *dependent variable* and the variables $x_1, x_2, \ldots, x_n$ are the *independent variables.* The $n + 1$ variables $x_1, x_2, \ldots, x_n, z$, together with the procedure which determines the values of z, constitute a *function.*

Caution: Following common usage, we shall frequently say that the dependent variable is the function, or that z is a function of $x_1, x_2, \ldots, x_n$.

We write $z = f(x_1, x_2, \ldots, x_n)$ to indicate that z is a function of the variables $x_1, x_2, \ldots, x_n$ and, if $a_1, a_2, \ldots, a_n$ are values for the variables $x_1, x_2, \ldots, x_n$ respectively, we denote by $f(a_1, a_2, \ldots, a_n)$ the unique value of z which is determined when the independent variables are given these values. We say that $f(a_1, a_2, \ldots, a_n)$ is the value of z which *corresponds* to the values $a_1, a_2, \ldots, a_n$ for $x_1, x_2, \ldots, x_n$ respectively.

According to Definition 3.3, a function may be defined by describing a procedure which determines a unique value of the dependent variable for each possible set of values of the independent variables. There are three main methods for describing such a procedure which we shall use at various times. We may give an equation which expresses the value $f(a_1, a_2, \ldots, a_n)$ in terms of the values $a_1, a_2, \ldots, a_n$; we may give a table which lists all possible sets of values of the independent variables and gives the corresponding value of the dependent variable in each case; we may give a set of rules by which a unique value of z may be determined when values for $x_1, x_2, \ldots, x_n$ are given.

There are several other noteworthy features of Definition 3.3; they are mentioned here and illustrated in Examples 3.4–3.6. It is not necessary that a change in the values of the independent variables produce a change in the corresponding value of the dependent variable; in fact, it may even happen that all possible values for the x's determine the same value of z. Moreover, there may be some elements in the range of z which do not correspond to any set of values of the independent variables. The only requirement is that for each possible set of values of the independent variables, a unique value of the dependent variable should be determined.

There is a special class of functions of one independent variable which is particularly important. A function $z = f(x)$ is said to be a *one-to-one correspondence* between the ranges of x and z iff, in addition to the requirements of Definition 3.3, every element in the range of z corresponds to exactly one element in the range of x. The concept of a one-to-one correspondence will be used in Section 3-3 to define isomorphism.

Example 3.4 Let x and z be variables each having the set of all real numbers as its range. For each real number r, set $f(r) = r^2$. This equation gives a procedure by which a value of z is determined whenever a value of x is given; thus z is a function of x. This procedure can be described more conveniently by the equation $z = x^2$. We shall use this more convenient notation; some practice with it is supplied in the problems at the end of this section. The equation appears to be saying something about the variables x and z but the interpretation is that the indicated operations should be performed on each of the values of x and the result is the corresponding value of z. With this function a change in the value of x does not necessarily produce a change in the corresponding value of z since $f(1) = f(-1)$. Thus, the function f is not a one-to-one correspondence. Not all values of z correspond to a value of x; for example, no value of x will produce the number -1 as the corresponding value of z.

x	z
1	0
2	0
3	0

FIGURE 3.1

Example 3.5 Let x be a variable with range $\{1, 2, 3\}$ and let z be a variable with range $\{0, 1\}$. Figure 3.1 gives a procedure by which a value of z is determined whenever a value of x is given. (Merely read across the appropriate row.) Thus z is a function of x. In this example, every value of x determines the same corresponding value of z.

Example 3.6 Let x, y, and z be sentential variables each having the set of all sentences as its range. To make z a function of x and y (in symbols $z = f(x, y)$) we may prescribe the following set of rules.

(1) If a is a true sentence, $f(a, b) = a \vee b$.

(2) If a is a false sentence, $f(a, b) = a \wedge b$.

A *sentence-valued function* is a function in which the dependent variable is a sentential variable; a *sentential function* is a function in which all the variables (both dependent and independent) are sentential variables Evidently every sentential function is a sentence-valued function but not conversely. For example, let x be a real variable, let z be a sentential variable, and set

$$z = (x \text{ is an integer}).$$

(For each value "a" of x, the corresponding value of z is the sentence "a is an integer.") This defines z as a sentence-valued function of x, but it is not a sentential function.

A sentential function is defined in Example 3.6. Each of our basic connectives also gives rise to a sentential function. In fact, each of the truth tables in Figs. 2.2–2.5 is really concerned with a sentential function and not with a particular sentence since different truth values are considered. For example the table in Fig. 2.2 (for "$a \wedge b$") deals with the function $z = x \wedge y$; similarly for the other connectives. By successive use of the truth tables for the basic connectives, truth tables can be constructed for more complicated functions which are defined in terms of these connectives. The procedure is illustrated in Example 3.7.

Example 3.7 A truth table for the function

$$z = x \to [y \wedge (x \vee y)]$$

is shown in Fig. 3.2. The two columns at the left list all of the possibilities for the pair of truth values of the sentences used as values of x and y. The numbers above the other columns give the order in which they were filled in. Columns 1, 2, and 3 were filled in using Figs. 2.3, 2.2, and 2.4 respectively. The final column, 3, gives the truth value of the sentence obtained as the corresponding value of z.

PROBLEMS

1. Let x, y, and z be variables each having the set of all real numbers as its range. In which of the following cases is z a function of x and y ?

(a) $z = x + y$.

(b) $z = x - y$.

(c) $z = x \cdot y$.

(d) $z = x/y$.

(e) $z = x$.

(f) $z = \begin{cases} x \text{ if } x > 0. \\ y \text{ if } x < 0. \end{cases}$

(g) $z = \begin{cases} 2x \text{ if } x > 0. \\ 3x \text{ if } y > 0. \end{cases}$

(h) $z = \begin{cases} 2 \text{ if } x^2 + y^2 > 0. \\ 1 \text{ if } x^2 + y^2 = 0. \\ -3 \text{ if } x^2 + y^2 < 0. \end{cases}$

(i) $z = (x + y)^{1/2}$.

2. Let x be a variable whose range is the set of all chairs in the Radio City Music Hall and let z be a variable whose range is the set of all human

FIGURE 3.2

		3	2	1
x	y	$x \to$	$[y \wedge$	$(x \vee y)]$
1	1	1	1	1
1	0	0	0	1
0	1	1	1	1
0	0	1	0	0

beings. In each of the following cases, information is given about $f(c)$ for each chair c in the range of x. In which cases is z a function of x?

(a) $f(c)$ is the person sitting in c.

(b) $f(c)$ is the person sitting in c at noon on Jan. 1, 1963.

(c) $f(c)$ is the last person to sit in c prior to noon on Jan. 1, 1963.

(d) $f(c)$ is the first person to sit in c after noon on Jan. 1, 1963.

(e) $f(c)$ is the person sitting closest to c at noon on Jan. 1, 1963.

(f) $$f(c) = \begin{cases} \text{John Jones, if } c \text{ is an aisle seat.} \\ \text{Mary Smith, if } c \text{ is not an aisle seat.} \end{cases}$$

(g) $$f(c) = \begin{cases} \text{The President of the United States, if } c \text{ is in the first row.} \\ \text{The Vice President of the United States, if } c \text{ is an aisle seat.} \\ \text{The Mayor of New York City, if } c \text{ is neither an aisle seat nor in the first row.} \end{cases}$$

3. Let x, y, and z be sentential variables each having the set of all sentences as its range. In which of the following cases is z a function of x and y?

(a) $z = x \vee y$.

(b) $z = \sim x$.

(c) $z = x \rightarrow y$.

(d) $z = (y \wedge \sim y) \vee (\sim x \wedge y) \vee (\sim x \wedge \sim y)$.

(e) $$z = \begin{cases} x, \text{ if } y \text{ is true.} \\ x \wedge y, \text{ if } y \text{ is false.} \end{cases}$$

(f) $$z = \begin{cases} x \leftrightarrow y, \text{ if } x \text{ and } y \text{ are both true.} \\ y, \text{ if } x \text{ and } y \text{ are both false.} \end{cases}$$

(g) $$z = \begin{cases} x \vee y, \text{ if not both of } x \text{ and } y \text{ are true.} \\ x \vee y, \text{ if not both of } x \text{ and } y \text{ are false.} \end{cases}$$

(h) $$z = \begin{cases} x \wedge \sim x, \text{ if } x \text{ and } y \text{ are both true.} \\ x \vee \sim x, \text{ if } x \text{ is true and } y \text{ is false.} \\ x \rightarrow y, \text{ if } x \text{ is false and } y \text{ is true.} \\ 4 = 5, \text{ if } x \text{ and } y \text{ are both false.} \end{cases}$$

4. Let each of x and y be a sentential variable whose range is the set of all sentences of the form "$p \cdot q = r$" where p, q, and r are positive integers. We shall use "$p_x \cdot q_x = r_x$" for a value of x; similarly, "$p_y \cdot q_y = r_y$" denotes a value of y. Let z be a sentential variable whose range is the set

$$\{\text{New York is a city in U.S.A.}, 2 \cdot 2 = 5, \sin \pi = -1\}.$$

In which of the following cases is z a sentential function of x and y?

(a) $z = \begin{cases} 2\cdot 2 = 5, \text{ if } x \text{ is true.} \\ \sin\pi = -1, \text{ if } x \text{ is false.} \end{cases}$

(b) $z = \begin{cases} \text{New York is a city in U.S.A., if } r_x < r_y. \\ 2\cdot 2 = 5, \text{ if } r_x = r_y. \\ \sin\pi = -1, \text{ if } r_x > r_y. \end{cases}$

(c) $z = \begin{cases} 2\cdot 2 = 5, \text{ if } p_x < p_y. \\ \sin\pi = -1, \text{ if } q_x < q_y. \end{cases}$

(d) $z = \begin{cases} \text{New York is a city in U.S.A., if } p_x\, q_x = p_y\, q_y. \\ 2\cdot 2 = 5, \text{ if } p_x\, q_x \neq p_y q_y. \end{cases}$

(e) $z = \begin{cases} 2\cdot 2 = 5, \text{ if } p_x\, q_x\, r_x = 2. \\ \sin\pi = -1, \text{ if } p_x\, q_x\, r_x \neq 2. \end{cases}$

(f) $z = \begin{cases} \text{New York is a city in U.S.A., if } r_x \leq r_y. \\ 2\cdot 2 = 5, \text{ if } r_x > r_y. \end{cases}$

(g) $z = \begin{cases} \text{New York is a city in U.S.A., if both } x \text{ and } y \text{ are true.} \\ 2\cdot 2 = 5, \text{ if one of } x \text{ and } y \text{ is true and the other is false.} \\ \sin\pi = -1, \text{ if both of } x \text{ and } y \text{ are false.} \end{cases}$

5. Let x and y be variables each having the set of all real numbers as its range. Let z be a sentential variable with the set of all sentences as its range. In which of the following cases is z a sentence-valued function of x and y?

(a) For each pair of values a, b of x and y respectively, the corresponding value of z is $a = b$. Using the convention mentioned in Example 3.4, this function can be described by the somewhat confusing notation $z = (x = y)$.

(b) $z = x > y$.

(c) $z = 2x + y$.

(d) $z = (xy \text{ is rational})$.

(e) $z = (x \text{ is a shaggy dog})$.

(f) $z \geq 2x - y$.

#6. Make a truth table for each of the following sentential functions.

(a) $z = x \rightarrow y$.

(b) $z = \sim x \vee y$.

(c) $z = \sim y \rightarrow \sim x$.

(d) $z = x \wedge \sim y \rightarrow \sim x$.

(e) $z = x \wedge \sim y \rightarrow y$.

*(f) $z = x_1 \wedge \sim x_2 \rightarrow x_3 \wedge \sim x_3$.

7. Make a truth table for each of the following sentential functions.

(a) $z = x \leftrightarrow y$.

(b) $z = (x \wedge y) \vee (\sim x \wedge \sim y)$.

(c) $z = (x \wedge y) \vee x$.

(d) $z = x \wedge (y \vee x)$.

(e) $z = x \rightarrow (y \wedge \sim x)$.

(f) $z = (x \rightarrow y) \wedge \sim x$.

8. Why is it possible to make a truth table for some sentential functions and not for others? Give an example of a sentential function for which it is not possible to make a truth table.

1-4 Related Sentential Functions; Truth Value Functions

We have seen in Section 1-2 that merely writing a sentence such as "$\sim a \vee \sim b$" gives us no information at all about the individual sentences "a" and "b." If we are told that this sentence is true we still cannot draw any conclusion about the sentences "a" and "b" separately. However, the truth of this sentence does tell us something about the pair of sentences "a" and "b" taken together; it tells us that not both of them have truth value 1. Two sentences which do not both have truth value 1 are said to be *inconsistent* with each other.

If two sentences "a" and "b" are such that the sentence "$a \rightarrow b$" is true, we say that "a" *implies* "b" and write "$a \Rightarrow b$". If "$a \leftrightarrow b$" is true, we say that "a" is *equivalent* to "b" and write "$a \Leftrightarrow b$". Using truth values we extend this notation a little farther and indicate that a sentence "a" is true by writing "$a \Leftrightarrow 1$"; similarly, "$a \Leftrightarrow 0$" tells us that the sentence "a" is false.

It is important to understand clearly the differences between "$a \rightarrow b$" and "$a \Rightarrow b$". The first sentence, "$a \rightarrow b$", is saying something about the things which are mentioned in the separate sentences "a" and "b". The statement it is making may be true, or it may be false. This sentence is one of the objects which are studied in logic; it belongs to our object language. On the other hand, "$a \Rightarrow b$" is saying something about the sentences "a" and "b"; moreover, it tells us that the statement it makes is true. It is a piece of information; it is not one of the objects which we study; it belongs to our metalanguage. As mentioned in Section 1-2, we shall frequently use the notation "if . . . , then . . .", when talking about sentences, to indicate that the statement made is actually true; that is, in the sense of "implies". In most cases the context will indicate whether we are concerned with an object for study or whether we are being given some information; in doubtful cases, we now have two different symbols and two different phrasings to distinguish between the two alternatives. Similar remarks apply to "$a \leftrightarrow b$" and "$a \Leftrightarrow b$".

The concepts of inconsistency, implication, and equivalence become more fruitful when generalized to apply to sentence-valued functions.

Let $u = f(x_1, x_2, \ldots, x_n)$ and $v = g(x_1, x_2, \ldots, x_n)$ be two sentence-valued functions of the same independent variables. Any one of

$$u \rightarrow v, \ f(x_1, x_2, \ldots, x_n) \rightarrow g(x_1, x_2, \ldots, x_n), \ f \rightarrow g$$

is defined to be a sentence which states that, for each set of values $a_1, a_2, \ldots, a_n$ of the independent variables, the sentence

$$f(a_1, a_2, \ldots, a_n) \rightarrow g(a_1, a_2, \ldots, a_n)$$

is true. Notice that "$f \rightarrow g$" is a sentence — not necessarily a true sentence; the statement that it makes is not asserted to be true merely because we write it down. Similarly, any one of

$$u \leftrightarrow v, \ f(x_1, x_2, \ldots, x_n) \leftrightarrow g(x_1, x_2, \ldots, x_n), \ f \leftrightarrow g$$

is a sentence (true or false) which states that

$$f(a_1, a_2, \ldots, a_n) \leftrightarrow g(a_1, a_2, \ldots, a_n)$$

is true for all sets of values $a_1, a_2, \ldots, a_n$ of the independent variables.

DEFINITION 4.1 Let $u = f(x_1, x_2, \ldots, x_n)$ and $v = g(x_1, x_2, \ldots, x_n)$ be two sentence-valued functions of the same independent variables. The functions f and g are *inconsistent* with each other iff, for each set of values $a_1, a_2, \ldots, a_n$ of the independent variables,

$$\sim [f(a_1, a_2, \ldots, a_n) \wedge g(a_1, a_2, \ldots, a_n)] \Leftrightarrow 1.$$

The function f *implies* the function g, in symbols

$$f(x_1, x_2, \ldots, x_n) \Rightarrow g(x_1, x_2, \ldots, x_n) \quad \text{or} \quad f \Rightarrow g \quad \text{iff } (f \rightarrow g) \Leftrightarrow 1.$$

The function f is *equivalent* to g, in symbols

$$f(x_1, x_2, \ldots, x_n) \Leftrightarrow g(x_1, x_2, \ldots, x_n) \quad \text{or} \quad f \Leftrightarrow g \quad \text{iff } (f \leftrightarrow g) \Leftrightarrow 1.$$

According to Definition 4.1, two sentence-valued functions f and g are inconsistent iff it never happens (that is, not for any set of values of the independent variables) that both of the values of f and g have truth value 1; f implies g iff it never happens that the value of f is true while the value of g is false; f is equivalent to g iff it never happens that the values of f and g have different truth values. Thus, in any one of these cases, there is at least a tenuous connection between the values of f and g; the functions are said to be *related*. We shall frequently be interested in the special case of sentential functions.

Example 4.2 Suppose we have a sack which contains a number of red squares and a number of blue circles; let each of x and y be a variable whose range is the set of objects in the sack. We define three sentence-valued functions f, g, and h by setting

$$f(x, y) = (x \text{ is red and } y \text{ is blue}).$$
$$g(x, y) = (x \text{ and } y \text{ have the same shape}).$$
$$h(x, y) = (x \text{ is red or } y \text{ is blue}).$$

The functions f and g are inconsistent; f implies h and also $g \Rightarrow h$.

Example 4.3 Let each of x and y be a sentential variable whose range is the set of all sentences. We define three sentential functions f, g, and h by setting

$$f(x, y) = \sim x \vee y.$$
$$g(x, y) = x \rightarrow y.$$
$$h(x, y) = x \leftrightarrow y.$$

The functions f and g are equivalent and the function h implies each of them.

In the definition of a sentential function there is no requirement that the truth value of the sentence obtained as a value of the dependent variable should be correlated in any way with the truth values of the sentences used as values for the independent variables. A particular correlation which is frequently desirable is taken, in the following definition, as the defining characteristic of a truth value function.

DEFINITION 4.4 A *truth value function* is a sentential function $f(x_1, x_2, \ldots, x_n)$ such that whenever $a_1, a_2, \ldots, a_n$ and $b_1, b_2, \ldots, b_n$ are in the respective ranges of $x_1, x_2, \ldots, x_n$ and $a_1 \Leftrightarrow b_1, a_2 \Leftrightarrow b_2, \ldots, a_n \Leftrightarrow b_n$ it follows that

$$f(a_1, a_2, \ldots, a_n) \Leftrightarrow f(b_1, b_2, \ldots, b_n).$$

Example 4.5 Let x and y be the usual sentential variables and set

$$f(x, y) = \begin{cases} x \vee y, & \text{if } x \text{ is true.} \\ x \wedge y, & \text{if } x \text{ is false.} \end{cases}$$

$$g(x, y) = \begin{cases} x, & \text{if } x \text{ mentions communism.} \\ y, & \text{if } x \text{ does not mention communism.} \end{cases}$$

The function f is a truth value function; g is not.

Truth tables provide a convenient procedure for showing that two truth value functions are equivalent. For example, the truth table for the truth value function $z = f(x, y)$ would list all possible pairs of truth values of sentences in the ranges of x and y (there are 4 such pairs) and, for each of the possibilities, the truth value of the corresponding sentence obtained for z would be given. For a truth value function $u = g(x, y)$ to be equivalent to f it is necessary and sufficient that, in each possible case, the truth values of the sentences obtained for z and u be the same. Thus it is necessary and sufficient that the final columns of the truth tables for the two functions be identical.

Example 4.6 Let x and y be the usual sentential variables, then

$$x \rightarrow y \Leftrightarrow \sim x \vee y.$$

Proof. Truth tables for $x \rightarrow y$ and for $\sim x \vee y$ were found in Problems 6a, b, Section 1-3. (The student should make these truth tables if he has not already done so.) The equivalence is evident since the truth tables are identical.

For many purposes a sentential function may be substituted for any equivalent one. That is, if two sentential functions are equivalent, it is immaterial which of the two is used. Under these circumstances, Example 4.6 shows that the connective "$\rightarrow$" could be dispensed with since any sentential function involving "$\rightarrow$" is equivalent to a sentential function in which "$\rightarrow$" has been replaced by certain occurrences of "$\sim$" and "$\vee$". Problem 5 shows that our other two connectives "$\wedge$" and "$\leftrightarrow$" are similarly expendable.

PROBLEMS

1. In Problem 4 of Section 1-3 you should have found that z is a sentential function of x and y in cases (a), (b), (d), (e), (f), and (g).

   (a) Which of these functions are truth value functions?

   (b) Which pairs of these functions are inconsistent?

   (c) What implications are there between two of these functions?

   (d) What equivalences are there between two of these functions?

2. Suppose that $f(x_1, x_2, \ldots, x_n)$ and $g(y_1, y_2, \ldots, y_m)$ are sentence-valued functions of (possibly) different independent variables. Try to think of some reasonable way of defining "$f \Rightarrow g$" and "$f \Leftrightarrow g$". [*Hint:* First consider the special case of $f(x)$ and $g(x, y)$.]

3. For the functions defined in Examples 4.2, 4.3, and 4.5, explain which are the dependent and which are the independent variables.

4. Let each of x and z be a sentential variable whose range is the set

$$\{1 + 1 = 2, 2 + 2 = 4, 3 + 3 = 3\}.$$

(a) How many different functions can be defined in which z is a function of x? (Two functions $f(x)$ and $g(x)$ are *equal* iff, for every value a of x, $f(a)$ and $g(a)$ are the same object.)

(b) How many of the functions in (a) are truth value functions?

(c) From the functions in (a), find several examples of pairs of (1) inconsistent functions, (2) equivalent functions, (3) functions, one of which implies the other.

5. Prove each of the following equivalences. The first three of these results show that, if a sentential function can be substituted for an equivalent one, then we could dispense with the three connectives "$\wedge$", "$\rightarrow$", and "$\leftrightarrow$".

(a) $x \wedge y \Leftrightarrow \sim(\sim x \vee \sim y)$.

(b) $x \leftrightarrow y \Leftrightarrow (x \wedge y) \vee (\sim x \wedge \sim y)$.

(c) $x \rightarrow y \Leftrightarrow \sim x \vee y$.

(d) $x \rightarrow y \Leftrightarrow x \wedge y \leftrightarrow x$.

(e) $x \rightarrow y \Leftrightarrow x \vee y \leftrightarrow y$.

#6. Prove that the following truth value functions are all equivalent. (That is, every two of them are equivalent.)

(a) $z = x \rightarrow y$.

(b) $z = \sim y \rightarrow \sim x$.

(c) $z = x \wedge \sim y \rightarrow y$.

(d) $z = x \wedge \sim y \rightarrow \sim x$.

(e) $z = x \wedge \sim y \rightarrow c \wedge \sim c$. (Here c is a sentence.)

7. In what circumstances would truth value functions be more useful than general sentential functions?

1-5 Some Methods of Proof

In mathematics, a *theorem* is a sentence which is true. A *proof* of the theorem is a proof that the sentence really is true. If we want to prove that a particular sentence is true it is evidently sufficient to prove that any equivalent sentence is true since, if one of two equivalent sentences is true, the other must also be true. Thus, if we have a sentence which we wish to prove is a theorem, it would be convenient to have several sentences which are equivalent to the given one in the hope that one of them would be easier to prove true. Many mathematical theorems are phrased in the form of an implication and we turn our attention to this case. Suppose the theorem is $a \Rightarrow b$. Since we are trying to prove the theorem, we certainly do not know whether the sentence "$a \rightarrow b$" is true or false; how could we find another sentence equivalent to this one without knowing its truth value? The equivalence of the two sentences must rest merely on the form in which they are written and not on the particular statement which is being made. Fortunately we are already acquainted with this situation. Problem 6 of Section 1-4 lists several equivalent truth value functions. Since these functions are equivalent we may choose any values for the independent variables and the sentences we obtain as the corresponding values of the dependent variables in the various functions will all be equivalent. Thus we find that, for any sentences a and b, the following sentences are all equivalent.

(1) $a \rightarrow b$.

(2) $\sim b \rightarrow \sim a$.

(3) $a \wedge \sim b \rightarrow b$.

(4) $a \wedge \sim b \rightarrow \sim a$.

(5) $a \wedge \sim b \rightarrow c \wedge \sim c$.

Thus we have a choice among 5 conditional statements. A proof that any one of them is true is a proof that all of them are true. We turn now to a short discussion of what is involved in a proof that a conditional statement is true.

In order that our discussion should be applicable to any one of the statements 1-5 above, we consider the conditional "$p \rightarrow q$". Moreover, we consider in turn each of the possible truth values for "p".

Case 1. The truth value of "p" is 0. In this case "$p \rightarrow q$" is true and we have nothing to prove.

Case 2. The truth value of "p" is 1. In this case we must prove that "q" is true.

Since in Case 1 there is nothing to prove, it is evident that it suffices to assume that "p" is true and prove that "q" is true. We shall not discuss the basis of validity of individual steps in a proof. Two examples are given below and further examples are supplied by proofs of later theorems.

Returning now to our choice among the 5 conditionals 1 through 5 listed above, a proof based on the form "$a \rightarrow b$" (i.e. a proof which assumes that "a" is true and proves that "b" is true) is called a *direct proof* of the theorem "$a \Rightarrow b$". A proof based on any of the other forms is called an *indirect proof* of the theorem "$a \Rightarrow b$". Various special names are sometimes used to refer to proofs based on a particular one of the forms 2, 3, 4, 5.

Example 5.1 Give a direct proof of the following theorem:† If n is an odd integer then n^2 is an odd integer.

Proof. We assume that n is an odd integer. Then $n - 1$ is an even integer so that $n - 1 = 2k$, and

$$n = 2k + 1 \quad \text{where } k \text{ is an integer.}$$

But then

$$n^2 = (2k+1)^2 = 4k^2 + 4k + 1 = 2(2k^2 + 2k) + 1$$

so that n^2 is an odd integer. ∎

Example 5.2 Give an indirect proof of the following theorem: If the square of an integer is even then the integer is even.

Proof. We base the proof on form 2 above. Hence we assume that the integer is not even and we must prove that its square is not even; but this was proved in Example 5.1. ∎

So far we have considered only methods for proving a theorem stated as an implication. There is another important method of proof which is applicable to a certain type of mathematical theorem. It is the method of mathematical induction; we shall assume some acquaintance with this method but shall describe the type of theorem to which it is applicable and shall briefly review the procedure for its use and the basis for its validity.

First a word of caution. Mathematical induction must not be confused with the process of "inductive reasoning" which is employed in the sciences. This latter process consists of examining many special cases of a certain type of event and then formulating sentences which it is hoped will be true in all cases of this type of event. The method has been exceedingly fruitful in scientific investigations, but it does not purport to be a rigorous proof. On the other hand, mathematical induction has an axiomatic basis and is a process for giving a rigorous proof for a certain type of theorem.

---

† This is the usual terminology; a more accurate phrasing would be: Give a direct proof that the following sentence is a theorem.

The type of theorem to which mathematical induction is applicable can be described as follows. Let n be a variable whose range is the set

$$N = \{1, 2, 3, \ldots\}$$

of all natural numbers (positive integers), and let $f(n)$ be a sentence-valued function of the independent variable n. The theorems in which we are interested are of the form

For every element n of N, "$f(n)$" is true.

That is, the theorem states that every one of a certain infinite collection of sentences is true. Moreover, these sentences are labelled by the positive integers.

The procedure in proving the theorem

For every element n of N, "$f(n)$" is true

by mathematical induction consists of two parts. Part 1 is to prove that the particular sentence "$f(1)$" is true. Part 2 is to prove $f(n) \Rightarrow f(n + 1)$; i.e. prove that for every element n of N, the sentence "$f(n) \rightarrow f(n + 1)$" is true. Before discussing why this two-part procedure is accepted as a proof of the theorem, we shall give an example.

Example 5.3 Prove that for every positive integer n

$$1 + 2 + 3 + \ldots + n = \frac{n(n + 1)}{2}.$$

Proof. PART 1. In this example, the particular sentence $f(1)$ is $1 = \frac{1 \cdot 2}{2}$, and this is well known to be true.

PART 2. In this example, the sentence $f(n)$ is

$$1 + 2 + 3 + \ldots + n = \frac{n(n + 1)}{2}$$

and the sentence $f(n + 1)$ is

$$1 + 2 + 3 + \ldots + n + (n + 1) = \frac{(n + 1)(n + 2)}{2}.$$

We must prove that the sentence $f(n)$ implies the sentence $f(n + 1)$ and the proof which we give must be valid for every positive integral value of n. We shall give a direct proof and we therefore assume "$f(n)$" is true; that is, we assume

$$1 + 2 + 3 + \ldots + n = \frac{n(n + 1)}{2}.$$

But, by adding $n + 1$ to each member of this equation, we find

$$1 + 2 + 3 + \ldots + n + (n + 1) = \frac{n(n + 1)}{2} + (n + 1) = \frac{(n + 1)(n + 2)}{2}$$

and this equation shows that "$f(n + 1)$" is true. ∎

The basis for the validity of a proof by mathematical induction lies in the axiom system describing the set N of natural numbers. We mention here only one of the axioms of this system; for a fuller discussion of the system see Ref. 11 or 18. The "axiom of induction" is an implication of the form $(a \wedge b \wedge c) \Rightarrow d$. In detail:

(0) S is a subset of N, and

(1) The number 1 is an element of S, and

(2) If a number n is an element of S then $n + 1$ is also an element of S

imply

$S = N$.

To see the connection between this axiom and a proof of the theorem

For every element n of N, "$f(n)$" is true

by mathematical induction, we let S be the subset of N composed of all the natural numbers n for which "$f(n)$" is true. Then the sentence 0 in the axiom is certainly true. Moreover, parts 1 and 2 of a proof by mathematical induction show, respectively, that the sentences 1 and 2 in the axiom are true. But then the axiom itself assures us that $S = N$, which proves the theorem.

PROBLEMS

Directions for Problems 1–10: Each of these problems gives an elementary theorem which is stated in the form of an implication. (Remember, "if . . . , then . . ." in a theorem has the force of "implies".) Prove each of these theorems and discuss your proof. Is it direct or indirect? If indirect, is it based on one of the conditionals 2 through 5 listed on p. 21 of the text? Which one?

1. If two triangles ABC and $A'B'C'$ have the three sides of one respectively equal to the three sides of the other, then the triangles are congruent.

2. If $x = 2$ then $x^2 - 5x + 6 = 0$.

3. If $x^2 - 5x + 6 = 0$, then $x \neq 5$.

4. If $x^2 - 5x + 6 = 0$, then $x < 5$.

5. If $AC = BC$ in the triangle ABC, then angle A equals angle B.

6. If angle A equals angle B in the triangle ABC, then $AC = BC$.

7. If two lines are cut by a transversal and a pair of alternate interior angles are equal, then the lines are parallel.

8. If $x > 5$ then $x^2 - 5x + 6 \neq 0$.

9. If $y = x^2$ and $1 < y < 4$, then $x < 2$.

10. If AB and CD are two distinct lines in a plane and each of these lines is perpendicular to a given line in that plane, then AB and CD are parallel.

11. Review the proofs of several theorems you know. Try to find a proof which is different from any of the types discussed in the text.

***12.** Look up a proof that $\sqrt{2}$ is irrational and discuss the proof. [A more general result is given in *The American Mathematical Monthly*, vol. 67 (1960) pp. 576-78.]

13. Use mathematical induction to prove each of the following theorems.

(a) For every element n of N

$$1 \cdot 2 + 2 \cdot 3 + \ldots + n(n+1) = \frac{n(n+1)(n+2)}{3} .$$

(b) For every element n of N, $2^n > n$.

(c) For every element n of N, if $x_1, x_2, \ldots, x_n$ are n distinct real numbers, then one of these numbers is the largest of them.

(d) If n is any integer larger than 3, then $2^n < n!$ Discuss the changes required in parts 1 and 2 of the proof because we are not considering all the natural numbers. What is the basis for the validity of the proof?

(e) If n is an integer larger than 1 then the maximum number of points of intersection of n lines in a plane is $\frac{1}{2}n(n-1)$.

14. Criticize the following "proof" by mathematical induction of the "theorem": For every element n of N, if $A_1, A_2, \ldots, A_n$ are n objects, then they are identical. (The notation of this section is used.)

PART 1. We must prove that any one object is the same as itself, and this is evidently true.

PART 2. We give a direct proof of the implication "$f(n) \Rightarrow f(n + 1)$". We therefore assume as true

$f(n)$: If $A_1, A_2, \ldots, A_n$ are any n objects, then they are identical

and we must prove that

$f(n + 1)$: If $A_1, A_2, \ldots, A_n, A_{n+1}$ are any $n + 1$ objects, then they are identical

is true. If we omit the object A_{n+1} we obtain n objects $A_1, A_2, \ldots, A_n$ which are all identical by $f(n)$ (which is assumed to be true). Now if we omit A_1, we also obtain n objects $A_2, A_3, \ldots, A_{n+1}$ which are all identical by $f(n)$ (since it applies to *any* n objects). But these two collections, each of n identical objects, overlap (for example A_2 is in both collections). Therefore all $n + 1$ objects are identical.

2

BOOLEAN FUNCTIONS

2-1 Introduction

Section 2-2 gives the definitions of Boolean constants, variables, and functions. Two tabular methods are presented for describing Boolean functions. Equivalence relations are defined in Section 2-3 and the connection between equivalence relations and decompositions is discussed. Section 2-4 uses the results of Sections 2-2 and 2-3 to establish a correspondence between truth value functions and Boolean functions.

2-2 Basic Definitions; Use of Tables

We have seen in Section 1-3 (Example 3.2) that a contant is a symbol which stands for one particular object. It is remarkable that an algebraic structure of considerable complexity can be built up using only two distinguishable objects. Moreover, this algebraic structure has wide applications in logic and in both pure and applied mathematics. The three definitions below define the basic concepts.

DEFINITION 2.1 A *Boolean constant* is one of the numbers 0 or 1.

It would be equally as acceptable to use any two different objects instead of the numbers 0 and 1. The particular choice we have made will be convenient because some (but unfortunately not all) of the manipulations we shall perform with Boolean constants will be naturally suggested by familiar properties of these numbers. We shall use α, β, γ, α_1, etc. as Boolean constants.

DEFINITION 2.2 A *Boolean variable* is a variable whose range is the set $\{0, 1\}$ of Boolean constants.

We shall use ξ, η, ζ, ξ_1, etc. for Boolean variables.

DEFINITION 2.3 A *Boolean function* is a function whose variables (both dependent and independent) are Boolean variables.

We shall use $\phi(\xi_1, \xi_2, \ldots, \xi_n)$, $\psi(\xi, \eta)$, Δ, etc. for Boolean functions.

Example 2.4 Each of the following equations defines a Boolean function. We are using the convention, explained in Example 3.4 of Section 1-3, of writing an expression involving variables to mean that the operations indicated in the expression should be performed with each of the values of the variables. The operations are the ordinary arithmetic ones of addition, multiplication, etc. The reader should prove that a Boolean function is defined in each case.

(a) $\phi(\xi, \eta) = \xi\eta$.

(b) $\phi(\xi_1, \xi_2, \ldots, \xi_n) = \max\{\xi_1, \xi_2, \ldots, \xi_n\}$.

(c) $\psi(\xi_1, \xi_2, \ldots, \xi_n) = \min\{\xi_1, \xi_2, \ldots, \xi_n\}$.

(d) $\Delta(\xi, \eta) = \xi + \eta - \xi\eta$.

(e) $\phi(\zeta_1, \zeta_2, \ldots, \zeta_n) = 1$.

(f) $\psi(\zeta_1, \zeta_2, \ldots, \xi_n) = 0$.

The use of equations to define Boolean functions, as in Example 2.4, has some disadvantages since it may be quite difficult to show that whenever each of the independent variables is assigned one of the values 0 or 1 the corresponding value of the dependent variable is sure to be either 0 or 1. Thus it may be hard to tell whether or not a particular expression is a satisfactory one to use in an equation defining a Boolean function.

ξ_1	ξ_2	ξ_3	ϕ
1	1	1	1
1	1	0	1
1	0	1	0
1	0	0	0
0	1	1	0
0	1	0	0
0	0	1	1
0	0	0	1

FIGURE 2.1

The use of a table to define a Boolean function avoids the difficulty mentioned above since it is very easy to inspect a particular table and tell whether or not it represents a Boolean function. Moreover, if only a few independent variables are involved, the tabular representation is concise and presents the relevant information in an easily accessible form. We shall usually limit ourselves to two or three independent variables in our examples. The tabular description of a general function was illustrated in Section 1-3 (Example 3.5); Example 2.5 below presents the notational conventions we shall use in connection with tabular descriptions of Boolean functions.

Example 2.5 A Boolean function ϕ of three independent variables ξ_1, ξ_2, ξ_3 is described in tabular form in Fig. 2.1. The first three columns in this table list all possible sets of values for the three independent variables; the last column gives the corresponding values of the dependent variable. We shall always list the sets of values of the independent variables in the same order so the columns dealing with the independent variables will be exactly the same for any functions of those variables. The only differences between the tables for two different functions of the same independent variables will be in the last column giving the values of the dependent variable.

A format different from that illustrated in Example 2.5 is sometimes more convenient for the tabular description of a Boolean function. This alternate format is described in Examples 2.6 and 2.7.

ξ \ η	1	0
1	0	1
0	1	1

FIGURE 2.2

ξ_1 \ ξ_2	0	0	1	1
1	0	0	1	1
0	1	1	0	0
ξ_3	0	1	0	1

FIGURE 2.3

Example 2.6 A table for a Boolean function of two independent variables ξ and η is shown in Fig. 2.2. The values of ξ are listed in the left hand column and those of η are listed in the top row of the table outside the guide lines. The entries in the body of the table give the values of the dependent variable corresponding to the 4 possible pairs of values for ξ and η.

Example 2.7 A table for a Boolean function of three independent variables is shown in Fig. 2.3. In fact, this table describes the function ϕ of Example 2.5. The values of the independent variables are again listed around the edges of the table outside the guide lines; values for ξ_1 are shown at the left, those for ξ_2 at the top, and for ξ_3 at the bottom. The entries in the body of the table give the values of the dependent variable corresponding to the different possible sets of values of the independent variables. The student should check that all the possible sets of values of the independent variables actually are represented in the table.

The proof of the following theorem shows how the format for the tabular representation of a Boolean function can be of help in obtaining general results on such functions.

ξ_1	ξ_2	$\cdots$	ξ_n	ϕ
1	1	$\cdots$	1	
1	1	$\cdots$	0	
$\vdots$	$\vdots$		$\vdots$	
0	0	$\cdots$	0	

FIGURE 2.4

THEOREM 2.8 There are $2^{(2^n)}$ different Boolean functions of n independent variables.

Proof. Figure 2.4 shows one of the standard formats for describing a Boolean function ϕ of the n variables $\xi_1, \xi_2, \ldots, \xi_n$. Two functions will be different if and only if the last columns in the tables for the two functions differ in at least one place. Thus the number of different functions is the number of different ways of filling in the last column.

First we find how many rows there are in the table. Since each of the n independent variables has 2 possible values and these are to be assigned in all possible ways, there are 2^n different sets of values for these variables and the table must have 2^n rows in order to list all of them.

Each position in the last column can be filled in 2 ways, by a "0" or by a "1". Thus, since there are 2^n rows, there are $2^{(2^n)}$ ways of filling in the last column, and this is the number of Boolean functions of n independent variables. ∎

PROBLEMS

1. Can a table similar to the ones in Figs. 2.2 and 2.3 be used for a Boolean function of 4 independent variables? What about 5 independent variables? n independent variables?
2. Make tables for the Boolean functions in Example 2.4. Which of the two tabular formats is most convenient?
3. (a) Find an equation which defines the Boolean function of Example 2.6.

   *(b) Find an equation which defines the Boolean function of Example 2.7.
4. Check the result of Theorem 2.8 by writing tables for all the Boolean

functions of 1 or 2 independent variables. Write the format for tables describing Boolean functions of 3 and 4 independent variables and check that these tables have the correct number of rows as given in the proof of Theorem 2.8.

5. Which of the following equations define Boolean functions?
   (a) $f(\xi, \eta) = \xi + \eta$.
   (b) $f(\xi, \eta) = \xi/\eta$.
   (c) $f(\xi, \eta) = \xi^\eta$.
   (d) $f(\xi, \eta, \zeta) = \xi + \eta + \zeta - \xi\eta - \eta\zeta$.
   (e) $f(\xi, \eta, \zeta) = \xi + \eta + \zeta - \xi\eta - \eta\zeta - \xi\zeta$.
6. For each of the conditions listed below, find the number of Boolean functions of n independent variables such that the final column in the table for the function (the column giving the values of the dependent variable) satisfies the condition.
   (a) There is a 1 in the first row.
   (b) All the entries are the same.
   (c) No two successive entries are the same.
   (d) There is exactly one 1.
   *(e) There are more ones than zeros.

2-3 Equivalence Relations

In this section we present the particularly important algebraic concept of an equivalence relation which we shall need in the next section to continue our study of Boolean functions. An equivalence relation, as the name implies, is characterized by abstracting the most useful properties of "equals". But first we must define a relation.

DEFINITION 3.1 A *binary relation* defined on a set S is a sentence-valued function of two independent variables each of which has S as its range.

We shall frequently refer to a binary relation simply as a relation. It would be possible to consider ternary relations as sentence-valued functions of three independent variables, etc., but we shall not have occasion to do so.

The notation used in connection with relations is somewhat different from that described in Section 1-3 in connection with general functions. We shall use the letter "R" to denote a relation (much as "f" has been used to denote a function) and shall use "$a\ R\ b$" instead of "$R(a, b)$" to denote the sentence which is the value of R when the independent variables have the values a and b respectively.

Example 3.2 (a) Let S be the set of all real numbers. For each pair of real numbers r and s, each of "$r = s$" and "$r < s$" is a sentence. Thus each of "$=$" and "$<$" is a sentence-valued function of two independent variables each having S as its range; that is, each of "$=$" and "$<$" is a binary relation defined on the set of all real numbers.

(b) Let S be the set of all residents of the United States. "Weighs more than" is a relation on S since, for each pair a, b, of residents of the United States, "a weighs more than b" is a sentence.

We are already acquainted with two relations defined on the set S of all sentence-valued functions with given independent variables. We have seen in Section 1-4 that if f and g are two such functions then each of "$f \rightarrow g$" and "$f \leftrightarrow g$" is a sentence. Thus, "$\rightarrow$" and "$\leftrightarrow$" are relations defined on S. These relations will be of considerable interest to us. We have already agreed to signify that the sentence "$f \rightarrow g$" is true by writing "$f \Rightarrow g$". Similarly, "$f \Leftrightarrow g$" means that the sentence "$f \leftrightarrow g$" is true. For an arbitrary relation R, we say that a is *related* to b (or *R-related* to b) in case the sentence "$a\ R\ b$" is true.

DEFINITION 3.3 An *equivalence relation* on a set S is a relation R defined on S and such that, for all elements a, b, c of S,

(1) a is R-related to a, and

(2) $a\ R\ b$ implies $b\ R\ a$, and

(3) $a\ R\ b$ and $b\ R\ c$ imply $a\ R\ c$.

The relation "$=$" on the set of all real numbers is an equivalence relation, as is also the relation "$\leftrightarrow$" on the set of all sentence-valued functions with given independent variables. The relation "$<$" on the set of all real numbers is not an equivalence relation; neither is the relation "$\rightarrow$" on the set of sentence-valued functions with given independent variables.

Each of the three conditions in Definition 3.3 has proved to be separately interesting and special names have been given to relations satisfying them individually. A relation is said to be *reflexive*, *symmetric*, or *transitive*, respectively, iff it satisfies condition 1, 2, or 3 of Definition 3.3. Thus an equivalence relation is a relation which is reflexive, symmetric, and transitive. The relation "$<$" on the real numbers is transitive, but neither reflexive nor symmetric.

Thinking of an equivalence relation R on a set S as a generalization of equality, we might hope to be able to divide S into subsets so that all of the elements of any one of the subsets are "the same as far as R is concerned". We would want the subsets to be *disjoint;* that is, the intersection of two different subsets should be empty. The next definition and theorem show that this is indeed possible.

DEFINITION 3.4 A *decomposition* of a set S is a collection of disjoint subsets of S whose union is S.

For example, if S is an ordinary xy-plane, the collection of all lines with zero slope is a decomposition of S. The collection containing the origin and all circles (circumferences) with center at the origin is also a decomposition of S, but the collection containing both the lines and the circles is not a decomposition since the subsets fail to be disjoint.

THEOREM 3.5 If R is an equivalence relation defined on a set S then there is a unique decomposition of S into non-empty subsets such that, for any elements a, b of S,

$$a\ R\ b \Leftrightarrow a \text{ and } b \text{ are in the same set of the decomposition.}$$

REMARK. The sets of the decomposition are called *equivalence classes* or *R-equivalence classes.*

Proof. It is easy to see that there cannot be two different decompositions satisfying the condition of the theorem. In fact, if two decompositions of S are different there are two elements a, b of S which lie in the same set of one of the decompositions and not in the other. But then both of the decompositions cannot satisfy the condition of the theorem since "$a\ R\ b$" is either definitely true or definitely false, whereas the sentence

a *and* **b** *are in the same set of the decomposition*

is true with respect to one of the decompositions and false with respect to the other. Thus, if there is a satisfactory decomposition it is unique.

To prove the existence of a satisfactory decomposition, for each element a of S denote by S_a the set of all elements b of S which are R-related to a; that is, S_a is the set of all elements b such that "$a\ R\ b$" is true. We show that the collection $\mathfrak{D}$ of all the sets S_a is a decomposition of S into non-empty sets. Certainly each of the sets S_a is a subset of S. Since R is reflexive, a is an element of S_a; this shows that each of the sets S_a is non-empty and also that the union of all of them is S. To show that these sets are disjoint, we assume that S_a and S_b are different but the element c is in both of them, and we derive a contradiction. Since S_a is different from S_b, one of these sets contains an element not in the other; say d is in S_a but not in S_b. Thus $a\,R\,d \Leftrightarrow 1$ and $b\,R\,d \Leftrightarrow 0$. Since c is in both S_a and S_b, both "$a\,R\,c$" and "$b\,R\,c$" are true. . The symmetric and transitive properties of R show $b\,R\,a \Leftrightarrow 1$, and another application of the transitivity gives $b\,R\,d \Leftrightarrow 1$. This contradiction completes the proof that $\mathfrak{D}$ is a decomposition of S into non-empty sets.

We have left to prove that $\mathfrak{D}$ satisfies the condition of the theorem. If "$a\,R\,b$" is true, then certainly a and b are in the same set of $\mathfrak{D}$ since both are in S_a. If both a and b are in the same set of $\mathfrak{D}$, say both are in S_c, then "$c\,R\,a$" and "$c\,R\,b$" are both true. But then the symmetric and transitive properties of R show that "$a\,R\,b$" is true. ▮

Theorem 3.5 shows that an equivalence relation on a set S gives rise to a decomposition of S. The following theorem shows that a decomposition also gives rise to an equivalence relation. Thus, these two concepts can be used interchangeably. Intuitively, an equivalence relation is concerned with a certain aspect or property of objects. Two objects are equivalent iff this particular aspect of the objects is the same, or iff their enjoyment of the special property is the same. An equivalence class is composed of all of the objects which are equivalent to each other.

THEOREM 3.6 If S is a set and $\mathfrak{D}$ is a decomposition of S into non-empty subsets, then there is a unique equivalence relation R defined on S such that, for all elements a, b of S

$$a\,R\,b \Leftrightarrow a \text{ and } b \text{ are in the same set of } \mathfrak{D}.$$

Proof. Problem 3. ▮

Example 3.7 (a) Let S be the set composed of all the stones in a certain pile and let R be the relation "weighs the same as" defined on S. Then R is an equivalence relation and the equivalence classes are the subsets of S containing all the stones of a particular weight. If our interest in the stones is to use them as weights on a balance scale, then all the stones in any one of the equivalence classes are exactly the same as far as our purposes are concerned. For some other purposes, for example, to use as a setting in a ring, the stones in one of the R-equivalence classes might vary widely in their desirability.

(b) Let S be the collection of all dimes and suppose Smith and Jones are two particular students. Form a decomposition of S into three equivalence classes by putting all the dimes in Smith's pocket in one equivalence class, all the dimes in Jones' pocket in a second, and all other dimes in a third. Then any two dimes in the same equivalence class are the same with respect to immediate availability for spending by Smith and Jones.

PROBLEMS

1. Which of the following are relations? Which are equivalence relations?

(**a**) "$\leftrightarrow$" on the set of all sentences; on the set of all sentence-valued functions with given independent variables. What about sentential functions or truth value functions?

(**b**) "$\Rightarrow$" on the set of all sentences, etc.

(**c**) "Has the same initials as" on the set of all students.

(**d**) "Is within 1 inch of being the same height as" on the set of all students.

(**e**) "Can beat up on" on the set of all children in Chicago.

(**f**) "Is the father of" on the set of all human beings.

(**g**) "Has been married to" on the set of all human beings.

(**h**) "Is married to" on the set of all human beings.

(**i**) "Can be depended upon" on the set of all policemen.

(**j**) "Is congruent to" on the set of all triangles.

2. (**a**) Find an example of a relation which is reflexive and symmetric but not transitive.

(**b**) Find an example of a relation which is transitive and reflexive, but not symmetric.

*(**c**) Find an example of a relation which is symmetric and transitive, but not reflexive.

3. (**a**) Prove Theorem 3.6.

(**b**) Discuss the logical structure of the proof given in the text for Theorem 3.5.

4. What are the equivalence classes of the equivalence relation "$\leftrightarrow$" on the set of all sentences? On the set of all truth value functions with $x_1, x_2, \ldots, x_n$ as independent variables?

5. Let R be an equivalence relation on a set S and let $\mathfrak{D}$ be the decomposition of S onto R-equivalence classes. According to Theorem 3.6, the decomposition $\mathfrak{D}$ gives rise to an equivalence relation on S; prove that this equivalence relation is R.

6. Let $\mathfrak{D}$ be a decomposition of a set S and let R be the unique equivalence relation on S which is connected with $\mathfrak{D}$ as in Theorem 3.6. According to Theorem 3.5, the equivalence relation R gives rise to a decomposition of S; prove that this decomposition is $\mathfrak{D}$.

7. Find several examples of equivalence relations which are familiar from every-day experience. Also find several examples of familiar decompositions of a set into equivalence classes. [*Hint:* These familiar situations may not have been previously looked at from this standpoint.]

#8. Let x and y be variables with ranges S and T respectively and such that y is a function of x; $y = f(x)$. Define "$a\ R\ b$" to mean $f(a) = f(b)$ and prove that R is an equivalence relation on S.

2-4 Correspondence between Truth Value Functions and Boolean Functions

In this section we present a connection between truth value functions and Boolean functions which will enable us to formulate many theorems of logic in terms of Boolean functions. The concept of an equivalence relation will be fundamental in our presentation.

Truth value functions and truth tables were mentioned briefly in Section 1-4; we shall need to take a closer look at them. Recall that a truth value function is a sentential function such that if two sets of values for the independent variables are composed of sentences which are respectively equivalent, then the two corresponding values of the dependent variable are equivalent sentences. This property is exactly what is required of a sentential function f in order that it have a truth table.

In fact, suppose the condition is satisfied, that is, suppose that f is a truth value function; then the truth values of the sentences which are substituted for the independent variables completely determine the truth value of the corresponding sentence obtained for the dependent variable. Thus, for each set $\alpha_1, \alpha_2, \ldots, \alpha_n$ of n Boolean constants, there is a Boolean constant β such that, whenever the sentences chosen as values for the respective independent variables have truth values $\alpha_1, \alpha_2, \ldots, \alpha_n$, the truth value of the corresponding sentence obtained for the dependent variable is β. Thus f has a truth table because such a table would list all the sets of n Boolean constants $\alpha_1, \alpha_2, \ldots, \alpha_n$ and give, for each set, the corresponding Boolean constant β.

Conversely, let f be a sentential function which is not a truth value function. Then there is a set $\alpha_1, \alpha_2, \ldots, \alpha_n$ of n Boolean constants and there are two different sets of n sentences, each having $\alpha_1, \alpha_2, \ldots, \alpha_n$ as their respective truth values, but such that the two sentences determined as the corresponding values of the dependent variable have different truth values. Thus there is no truth table for the function f because there isn't a unique Boolean constant to correspond to the set $\alpha_1, \alpha_2, \ldots, \alpha_n$.

It is evident that the truth table for a given truth value function is unique; this uniqueness is used in the following theorem to establish a correspondence between truth value functions and Boolean functions. We use a_1 etc. for sentences and x_1 etc. for sentential variables.

THEOREM 4.1 To each truth value function $f(x_1, x_2, \ldots, x_n)$ there corresponds a unique Boolean function $\phi(\xi_1, \xi_2, \ldots, \xi_n)$ such that

$$[(a_1 \leftrightarrow \alpha_1) \wedge (a_2 \leftrightarrow \alpha_2) \wedge \ldots \wedge (a_n \leftrightarrow \alpha_n)]$$

$$\Rightarrow [f(a_1, a_2, \ldots, a_n) \leftrightarrow \phi(\alpha_1, \alpha_2, \ldots, \alpha_n)].$$

Proof. The truth value function f has a unique truth table and this table itself determines a Boolean function ϕ with the desired property. Moreover, any Boolean function satisfying the condition of the theorem must have, for its table, the truth table of the function f. Thus the Boolean function ϕ is uniquely determined by f. ∎

Let u and v be variables whose ranges are respectively the set $\mathcal{T}_n$ of all truth value functions with n independent variables and the set $\mathcal{B}_n$ of all Boolean functions with n independent variables. According to Theorem 4.1, v is a function of u, say $v = F(u)$. But now by Problem 8, Section 2-3, we obtain an equivalence relation R on $\mathcal{T}_n$ by defining "$f\,R\,g$" to

FIGURE 4.1

Truth value function $f(x_1, x_2, \cdots, x_n)$	Boolean function $\phi(\xi_1, \xi_2, \cdots, \xi_n)$
x_i is a sentential variable value of x_i: a sentence, a_i value of f: a sentence	ξ_i is a Boolean variable value of ξ_i: 0 or 1, α_i value of ϕ: 0 or 1
a_i is $\begin{cases} \text{true} \\ \text{false} \end{cases}$	$\alpha_i = \begin{cases} 1 \\ 0 \end{cases}$
$f(a_1, a_2, \cdots, a_n)$ is $\begin{cases} \text{true} \\ \text{false} \end{cases}$	$\phi(\alpha_1, \alpha_2, \cdots, \alpha_n) = \begin{cases} 1 \\ 0 \end{cases}$

mean $F(f) = F(g)$. This equivalence relation R is just the relation "$\leftrightarrow$" with which we are already acquainted (Problem 5). Thus the R-equivalence classes are the sets of equivalent truth value functions of n independent variables. Two truth value functions in the same equivalence class correspond to the same Boolean function, but truth value functions in different equivalence classes correspond to different Boolean functions. A Boolean function may be thought of as representing an entire equivalence class of truth value functions (see Problem 4) and this is exactly what is needed in situations where equivalent truth value functions may be substituted for each other.

The correspondence set up by the function F will again prove to be convenient in the next chapter when we discuss implications. Figure 4.1 shows the principal features of this correspondence. The Boolean function which is identically equal to 1 is particularly important in this correspondence; a truth value function which corresponds to this Boolean function is called a *tautology*. We shall sometimes write "$f \leftrightarrow 1$" instead of the sentence: "f is a tautology." Similarly, "$f \leftrightarrow 0$" may be used instead of :"The Boolean function which corresponds to f is identically zero." [*Caution:* "$f \leftrightarrow 1$" is a sentence, but not necessarily a true sentence.]

Example 4.2 (a) The truth value function $x \wedge y$ has the truth table shown in Fig. 2.2, Section 1-2, and this table describes the Boolean function $\xi \cdot \eta$ which therefore corresponds to $x \wedge y$.

(b) The truth table for the truth value function $x \vee \sim x$ shows that its corresponding Boolean function is identically 1. Thus, $x \vee \sim x$ is a tautology.

PROBLEMS

1. For each of the following truth value functions, find the Boolean function which corresponds to it.
   (a) $\sim x$.
   (b) $x \vee y$.
   (c) $x \wedge y$.
   (d) $x \rightarrow y$.
   (e) $x \leftrightarrow y$.
2. For each of the truth value functions of Problem 4, Section 1-3, find the Boolean function which corresponds to it.

3. For each of the Boolean functions in Example 2.4, find a truth value function to which it corresponds. Is the truth value function unique in each case?

*4. For every Boolean function, is there a truth value function to which it corresponds?

5. For the equivalence relation R defined just after the proof of Theorem 4.1, prove that $f \, R \, g \Leftrightarrow f \leftrightarrow g$.

6. Find three tautologies.

3

ORDERED SETS

3-1 Introduction

Ordering relations are defined in Section 3-2 and a graphical representation of finite ordered sets is explained. In Section 3-3 an isomorphism between two general algebraic systems is defined and the definition is specialized to apply to ordered sets. In Section 3-4 an ordering relation is introduced in a natural way into the set $\mathcal{B}_n$ of all Boolean functions of n independent variables and in Section 3-5 this ordering relation is shown to be convenient for expressing certain results in logic.

3-2 Basic Definition; Graphical Representation

The concept of a relation was defined in Section 2-3 and the special case of an equivalence relation was discussed in some detail. In this section we shall be interested in another special type of relation.

DEFINITION 2.1 An *ordering relation* is a relation R defined on a set S such that for all elements a, b, c of S,

(1) $a\,R\,a \Leftrightarrow 1.$

(2) $(a\,R\,b) \wedge (b\,R\,a) \Rightarrow a = b.$

(3) $(a\,R\,b) \wedge (b\,R\,c) \Rightarrow a\,R\,c.$

It is instructive to compare these three conditions with those which characterize an equivalence relation. Conditions 1 and 3 are the familiar reflexive and transitive properties. Condition 2 almost contradicts the symmetry property; it says that if "$a \, R \, b$" is true, and "a" and "b" are different, then "$b \, R \, a$" is false. A relation satisfying Condition 2 is said to be *antisymmetric*. A word of caution should be added. There are several slight variations of our definition of an ordering relation which are in common use. The student is advised to check carefully to see exactly what definition is being used whenever the term "ordering" is introduced.

DEFINITION 2.2 An *ordered set* is a set S together with an ordering relation R on S. We say that S is *ordered* by the relation R.

Example 2.3 Each of the following examples is an ordered set.

(a) S is the set of all real numbers; for each pair a, b of real numbers, $a \, R \, b \Leftrightarrow a \leq b$.

(b) S is the set of all positive integers; for each pair a, b of real numbers "$a \, R \, b$" is defined as "a is a divisor of b".

(c) S is a collection of sets; for each pair A, B of sets in the collection S, "$A \, R \, B$" is defined as "A is a subset of B", in symbols, $A \subset B$. This relation is called the relation of *inclusion*.

Further examples will be found in the problems. The ordered set of Example 2.3a is a particularly important one; it serves as a prototype of all ordered sets in the sense that an ordering relation is defined by abstracting the most useful properties of "$\leq$" just as an equivalence relation was defined by abstracting the properties of "$=$". For this reason, the symbol "$\leq$" is frequently used to denote any ordering relation, and we shall follow this custom when convenient. Thus, when we denote a relation by the symbol "$\leq$", it is to be assumed that the relation is reflexive, antisymmetric, and transitive.

The ordered set of Example 2.3c will also be important in our future work, especially the case of the collection of all subsets of a given set. We shall frequently use a script letter for a collection of sets, so that the typography will indicate the complexity of the set under consideration. For example, given the set $S = \{0, 1\}$, we may consider the set $\mathcal{S}$ of all subsets of S. Then $\mathcal{S}$ is a set with 4 elements which is ordered by inclusion. The relation of inclusion between subsets of a given set is really a more typical ordering relation than is the relation "$\leq$" between real numbers. The reason is that every two real numbers a and b are *comparable;* that is

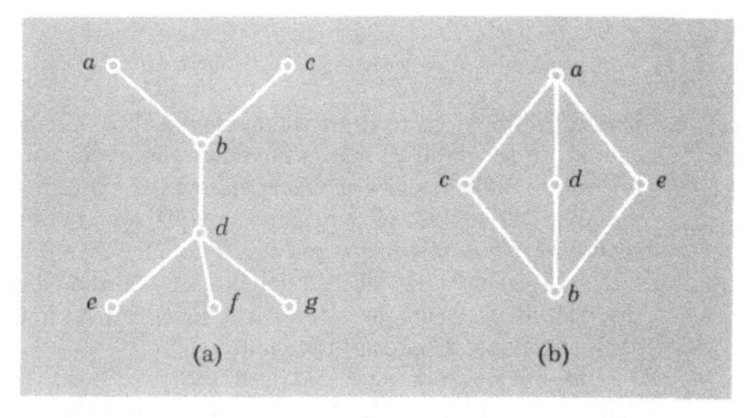

FIGURE 2.1

at least one of the sentences "$a \leq b$" and "$b \leq a$" must be true. On the other hand, it is easy to find two sets A and B such that both of the sentences "$A \subset B$" and "$B \subset A$" are false. Thus, for the ordering relation "$\subset$", it is possible to have an *incomparable* pair of elements; that is, there may be two elements such that neither one is related to the other.

Although an ordered set may be infinite, many of the examples in which we shall be interested will be finite. For finite ordered sets there is a system of graphical representation which is helpful in visualizing the ordering relation. Two examples are shown in Fig. 2.1. Figure 2.1a shows a set containing 7 elements. These elements are represented by the 7 small circles labelled $a, b \ldots, g$. The line segments in the figure indicate which pairs of elements are related by the ordering relation in the set; one element is related to another if and only if it is possible to go from the circle representing the first element to the circle representing the second element, moving along line segments in the figure, and always moving upward. For example, "$e \leq b$" is true since it is possible to move from e to d and then to b along upward sloping line segments. Similarly, "$f \leq a$" and "$b \leq c$" are each true. However, each of "$a \leq c$" and "$b \leq d$" is false since, in each case, it is impossible to move from the first element to the second along upward sloping line segments.

There may be some intersecting line segments in the graphical representation of a finite ordered set as illustrated in Fig. 2.2a. An alternative

FIGURE 2.2

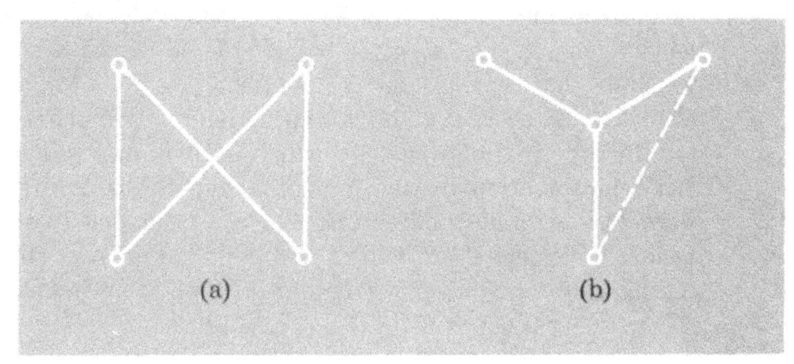

procedure would be to draw the figures in three-dimensional space, using curved arcs instead of line segments. Since intersecting line segments present no difficulty in visualizing an ordered set from its graphical representation, we shall draw our figures in a plane and use line segments. We shall also use as few line segments as possible in representing an ordered set. For example, the dashed line in Fig. 2.2b should not be put in, since the relation between the elements it joins can be found from the other line segments and the transitive law.

One obvious difference between the two illustrations in Fig. 2.1 is that (a) "spreads apart" at the top and bottom while (b) "comes together". This difference suggests the following definition.

DEFINITION 2.4 A *greatest* element of an ordered set S is an element G of S such that, for each element a of S, $a \leq G \Leftrightarrow 1$. A *least* element of an ordered set S is an element L of S such that, for each element a of S $L \leq a \Leftrightarrow 1$.

We shall reserve the capital letters "L" and "G" for the least and greatest elements, respectively, of an ordered set. The ordered set of Fig. 2.1a has neither a least nor a greatest element. For example, $a \neq G$ because "$c \leq a$" is false. The ordered set of Fig. 2.1b has both a greatest and a least element. It is easy to check that $a = G$ and $b = L$ in this set.

In using the graphical representation of an ordered set it is natural to inquire which ordered sets can be pictured in this way. Fortunately, any finite ordered set can be so represented, as the following theorem shows.

THEOREM 2.5 Any finite ordered set has a graphical representation (as illustrated in Fig. 2.1).

Proof. The proof is by induction on the number of elements in the set. Certainly any ordered set with one element is represented by a single small circle and no line segments. (The empty ordered set is represented by a blank picture.)

Now assume that any ordered set with n elements can be pictured and let $S = \{a_1, a_2, \ldots, a_n, a_{n+1}\}$ be an ordered set with $n + 1$ elements. If we omit the element a_{n+1} we obtain a set $S' = \{a_1, a_2, \ldots, a_n\}$ with n elements and there is an obvious ordering relation defined in S' by restricting the ordering relation "$\leq$" which is given in S. Thus S' has a graphical representation by the induction hypothesis, and it remains only to show that a circle representing a_{n+1} can be added to this representation. Define three sets A, B, and C as follows. Put in A all the elements x of S' such that "$x \leq a_{n+1}$" is true; put in B all the elements x of S' such that

"$a_{n+1} \leq x$" is true; put all the other elements of S' in C. It is easy to prove (Problem 7) that the three sets A, B, C form a decomposition of S'. Moreover, for each pair of elements a, b such that a is in A and b is in B, "$a \leq b$" is true since each of "$a \leq a_{n+1}$" and "$a_{n+1} \leq b$" is true. Thus, in the graphical representation of S', every circle representing a point in A is below each of the circles representing a point in B. The circle representing a_{n+1} may be drawn between these two collections of circles (above the circles representing points of A and below those representing points of B). A graphical representation of the ordered set S is obtained by drawing line segments downward from this new circle to the appropriate circles in A, and upward from the new circle to the appropriate circles in B. ∎

PROBLEMS

1. Which of the following relations are ordering relations in the respective sets?

   (a) S is the set of integers; "$a\ R\ b$" means "a is a divisor of b."

   (b) S is the set of natural numbers; "$a\ R\ b$" means "a has fewer prime factors than b."

   (c) S is the set of natural numbers; "$a\ R\ b$" means "a has no more prime factors than b."

   (d) S is the set of natural numbers; "$a\ R\ b$" means "a has fewer prime factors than b, or a and b have the same number of prime factors and a is less than or equal to b."

   (e) S is the set $\mathcal{B}_n$ of all Boolean functions of n independent variables; "$\phi\ R\ \psi$" means "for each set $\alpha_1, \alpha_2, \ldots, \alpha_n$ of n Boolean constants, $\phi(\alpha_1, \alpha_2, \ldots, \alpha_n)$ is less than or equal to $\psi(\alpha_1, \alpha_2, \ldots, \alpha_n)$."

   (f) S is the set $\mathcal{B}_n$; "$\phi\ R\ \psi$" means "for each set $\alpha_2, \alpha_3, \ldots, \alpha_n$ of $n - 1$ Boolean constants, $\phi(0, \alpha_2, \alpha_3, \ldots, \alpha_n)$ is less than or equal to $\psi(0, \alpha_2, \alpha_3, \ldots, \alpha_n)$."

   (g) S is the set $\mathcal{B}_n$; "$\phi\ R\ \psi$" means "for each set $\alpha_2, \alpha_3, \ldots, \alpha_n$ of $n - 1$ Boolean constants, $\phi(0, \alpha_2, \alpha_3, \ldots, \alpha_n)$ is less than or equal to $\psi(1, \alpha_2, \alpha_3, \ldots, \alpha_n)$."

   (h) S is the set $\mathcal{B}_n$; "$\phi\ R\ \psi$" means "$\phi(0, 0, \ldots, 0) \leq \psi(0, 0, \ldots, 0)$."

   (i) S is the set $\mathcal{B}_n$; "$\phi\ R\ \psi$" means "$\phi(0, 0, \ldots, 0) \leq \psi(1, 1, \ldots, 1)$."

2. (a) For each of the ordered sets you found in Problem 1b, c, d, con-

sider only the subset $S' = \{1, 2, 3, \ldots, 10\}$ of S and draw a graphical representation of S'.

(b) For each of the ordered sets you found in Problem 1e, f, g, h, i, draw a graphical representation of the ordered set in the case $n = 1$.

3. Which of the ordered sets you found in Problem 1 have a greatest element? Which have a least element?

4. Prove that each of the Examples 2.3a, b, c is an ordered set.

5. Can every set be ordered, or does there exist a set on which no ordering relation can be defined?

6. If S is a set which is ordered by the relation "$\leq$", prove that S is also ordered by the relation "$\geq$", where "$a \geq b$" means $b \leq a$.

7. (a) Prove that the three sets A, B, C described in the proof of Theorem 2.5 form a decomposition of S'.

(b) What are the "appropriate" circles in A and B as mentioned at the end of the proof of Theorem 2.5?

8. Prove that any ordered set contains at most one least element and at most one greatest element.

#9. For each of the following conditions, give two examples of an ordered set which satisfies the condition.

(a) There is no greatest element and there is no least element.

(b) There is a greatest element but there is no least element.

(c) There is a greatest element and also a least element.

10. If an ordered set does not contain a least element, can it always be enlarged in some way so that the enlarged ordered set does contain a least element?

#11. For the set $S = \{a, b, c\}$ with three distinct elements, how many relations are there which order S? (Two relations R and R' are different iff there is a pair of elements p, q of S such that one of "$p\ R\ q$" and "$p\ R'q$" is true and the other one is false.)

12. Is every finite collection of small circles and line segments the graphical representation of some ordered set? If not, can you characterize the pictures which do represent finite ordered sets?

#13. Let G be the greatest element in a set ordered by "$\leq$". Prove that

$$a = G \text{ and } a \leq b \Rightarrow b = G.$$

#14. Let S be any set and let $\mathfrak{S}$ be a collection of subsets of S such that the empty set $\emptyset$ and the set S are each in $\mathfrak{S}$. Define a relation "$\leq$" in the family of all subsets of S by

$$A \leq B \Leftrightarrow \text{There is an } X \text{ in } \mathfrak{S} \text{ such that } A \subset X \subset B.$$

Prove each of the following.

(a) $\emptyset \leq \emptyset \Leftrightarrow 1$.

(b) $S \leq S \Leftrightarrow 1$.

(c) $A \leq B \Rightarrow A \subset B$.

(d) $A \subset A_1 \leq B_1 \subset B \Rightarrow A \leq B$.

(e) The relation "$\leq$" is transitive.

(f) $\mathfrak{S} = \{A \mid A \subset S \text{ and } A \leq A\}$.

3-3 Isomorphism

We defined a relation in Section 2-3 and, in Section 3-2, we defined an ordered set as a set with a certain type of relation defined on it. We are frequently interested in things which can be done with two elements of a set as well as in relations between the elements.

DEFINITION 3.1 A *binary operation* defined on a set S is a function of two independent variables, in which all three of the variables (dependent and independent) have range S.

For example, addition is a binary operation defined on the set of real numbers. The term "binary operation" is frequently shortened to "operation"; it would be possible to consider ternary operations, etc. as certain functions of three or more independent variables, but we shall not do so. Just as with relations, the notation used in connection with operations is different from the functional notation described in Section 1-3. For the operation addition, "$+$", on the real numbers, we denote the value of the dependent variable which corresponds to the pair of values a, b for the independent variables by "$a + b$" instead of by "$+(a, b)$."

In an operation on S, both the dependent and independent variables have range S; however, it must be remembered that the dependent variable is treated differently from the independent variables. For an operation "$*$" defined on a set S it is required that, for each pair a, b of elements of S, $a * b$ must be an element of S. It is not required that each element of S be expressible in the form $a * b$. For example, we might define an operation "$\bigcirc$" on the real numbers by setting $a \bigcirc b = 0$ for each pair of real numbers a, b.

DEFINITION 3.2 An *algebraic system* is a set S together with certain relations or operations defined on S.

For example, an ordered set is an algebraic system, as is also the set of all complex numbers with the operations of subtraction and multiplication.

There are 19 different ordering relations in the set $S = \{a, b, c\}$ containing 3 distinct elements (Problem 11, Section 3-2). However, several of these ordering relations result in graphical representations which are the same except for the labelling of the small circles. Figure 3.1 shows the graphical representations of two different orderings of the set S. It is evident that the ordering of Fig. 3.1b is obtained by merely interchanging the labels "a" and "b" in the ordering of Fig. 3.1a. These two orderings are not the same since "$a \leq b$" is true in one of them and false in the other. Yet, Fig. 3.1 shows that there certainly is some kinship between the two orderings. This type of kinship is the subject of the next definition, but first we need one more bit of terminology.

We say that two algebraic systems are of the *same type* iff there is a one-to-one correspondence between the two sets of relations involved in the systems and also a one-to-one correspondence between their operations. We shall frequently use the same symbol to denote corresponding operations or relations in two algebraic systems of the same type. For example, any two ordered sets are algebraic systems of the same type since each has one relation and no operations.

DEFINITION 3.3 Two algebraic systems of the same type are *isomorphic* iff there is a one-to-one correspondence between the two sets which preserves corresponding relations and operations. Such a one-to-one correspondence is called an *isomorphism*.

The word "preserves" in Definition 3-3 needs some clarification. Suppose that (a, a'), (b, b'), and (c, c') are any three pairs of corresponding

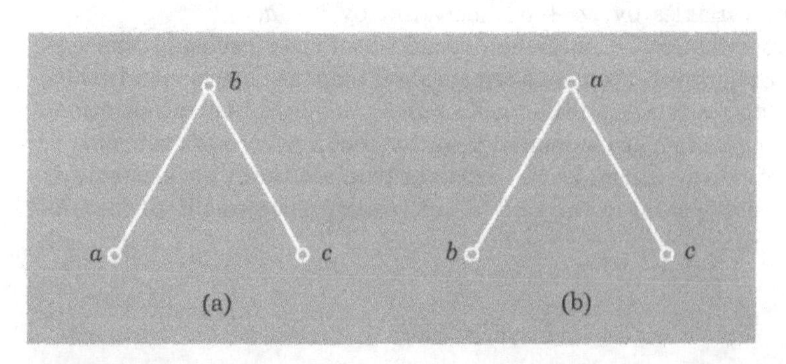

FIGURE 3.1

elements in the two algebraic systems, and that "R" and "$*$" are a relation and an operation, respectively, defined on both systems. (We are using the same symbol for corresponding relations, etc.) The requirement of Definition 3.3 is that

$$a \, R \, b \Leftrightarrow a' \, R \, b'$$

and

$$a * b = c \Leftrightarrow a' * b' = c'.$$

In particular, two ordered sets are isomorphic iff there is a one-to-one correspondence between the sets such that $a \leq b \Leftrightarrow a' \leq b'$ where (a, a') and (b, b') are any pairs of corresponding elements.

It is now easy to see that the ordered sets of Fig. 3.1a and Fig. 3.1b are isomorphic since the correspondence in which the pairs of corresponding elements are (a, b), (b, a), and (c, c) is an isomorphism between these two ordered sets.

PROBLEMS

1. (**a**) How many orderings are there in a set of 3 distinct elements, such that no two of the orderings are isomorphic?

(**b**) How many orderings are there in a set of 4 distinct elements, such that no two of the orderings are isomorphic?

2. Consider the algebraic system composed of the set $N = \{1, 2, 3, \ldots\}$ of all positive integers with the usual ordering. Show that there is only one isomorphism of this algebraic system with itself.

3. Consider the algebraic system composed of the set $I = \{\ldots, -2, -1, 0, 1, 2, \ldots\}$ of all integers with the usual ordering. Show that there are an infinite number of isomorphisms of this algebraic system with itself.

4. Consider the algebraic system composed of the set I of all integers with the usual ordering and with the operation of addition. How many isomorphisms are there of this algebraic system with itself?

5. Consider the algebraic system composed of the set I of all integers with the operation of addition. How many isomorphisms are there of this algebraic system with itself?

***6.** Consider the algebraic system composed of the set I of all integers with the operation of multiplication. Find two different isomorphisms of this algebraic system with itself. Show that there are an infinite number of such isomorphisms.

7. Find an ordered set which has a first element, and such that each element has an immediate successor and each element except the first has an immediate predecessor, but which is not isomorphic to the set $N = \{1, 2, 3, \ldots\}$ of all positive integers.

3-4 The Ordering in $\mathcal{B}_n$

In Chapter 2, we have already considered the set $\mathcal{B}_n$ of all Boolean functions of n independent variables. In this section, we shall define an ordering relation on $\mathcal{B}_n$, thus making it an algebraic system. In Chapter 4 we shall define two operations on $\mathcal{B}_n$ but, at present, we shall be interested in it only as an ordered set.

DEFINITION 4.1 A Boolean function $\phi(\xi_1, \xi_2, \ldots, \xi_n)$ is *less than or equal to* the Boolean function $\psi(\xi_1, \xi_2, \ldots, \xi_n)$, in symbols, $\phi \leq \psi$, iff for every set $\alpha_1, \alpha_2, \ldots, \alpha_n$ of n Boolean constants

$$\phi(\alpha_1, \alpha_2, \ldots, \alpha_n) \leq \psi(\alpha_1, \alpha_2, \ldots, \alpha_n) \Leftrightarrow 1.$$

Notice that the relation "$\leq$" which appears in the last line of Definition 4.1 is the usual ordering in the set $\{0, 1\}$ of Boolean constants (values of Boolean functions). This familiar ordering relation is used to define the ordering relation (also denoted by "$\leq$")on the set $\mathcal{B}_n$. Of course, it must be proved that this new relation actually is an ordering relation on $\mathcal{B}_n$.

THEOREM 4.2 The relation "$\leq$" defined in Definition 4.1 is an ordering relation on $\mathcal{B}_n$.

Proof. Problem 1 ∎.

The ordered set $\mathcal{B}_n$ has both a greatest and a least element. The greatest element is the function which is identically 1, and the least element is the function which is identically zero. In the next section we shall see that these two functions play a special role in the connection between logic and the ordered set $\mathcal{B}_n$.

Example 4.3 Each of the following statements is true of the ordering relation on $\mathcal{B}_2$.

$$\xi + \eta - \xi\eta \leq \max\{\xi, \eta\}.$$

$$\xi\eta \leq \max\{\xi, \eta\}.$$

PROBLEMS

1. Prove Theorem 4.2.

2. Find a graphical representation of the ordered set of all Boolean functions of one independent variable; of two independent variables.

3. For each of the following Boolean functions ψ, find how many different Boolean functions ϕ make the sentence "$\phi \leq \psi$" true.

   (a) $\psi(\xi, \eta, \zeta) = 1.$

   (b) $\psi(\xi, \eta, \zeta) = 0.$

   (c) $\psi(\xi, \eta, \zeta) = \xi.$

   (d) $\psi(\xi, \eta, \zeta) = \max \{\xi, \eta\}.$

   (e) $\psi(\xi, \eta, \zeta) = \xi\eta\zeta.$

4. For each of the following Boolean functions ϕ, find how many different Boolean functions ψ make the sentence "$\phi \leq \psi$" true.

   (a) $\phi(\xi, \eta, \zeta) = 1.$

   (b) $\phi(\xi, \eta, \zeta) = 0.$

   (c) $\phi(\xi, \eta, \zeta) = \xi.$

   (d) $\phi(\xi, \eta, \zeta) = \max \{\xi, \eta\}.$

   (e) $\phi(\xi, \eta, \zeta) = \xi\eta\zeta.$

5. Find two Boolean functions ϕ and ψ, with the same independent variables, such that both of "$\phi \leq \psi$" and "$\psi \leq \phi$" are false. These two functions are incomparable in the ordered set $\mathcal{B}_n$. What is the least number of independent variables in a pair of incomparable Boolean functions?

6. For each of the Boolean functions ψ of Problem 3, find how many Boolean functions of 3 independent variables are incomparable to ψ.

3-5 Connection with Logic

The correspondence between truth value functions and Boolean functions described in Section 2-4 enables us to write an implication between truth value functions of n independent variables in terms of the ordering relation in $\mathcal{B}_n$.

THEOREM 5.1 The truth value function $f(x_1, x_2, \ldots, x_n)$ implies the truth value function $g(x_1, x_2, \ldots, x_n)$ if and only if for their respective corresponding Boolean functions ϕ and ψ, "$\phi \leq \psi$" is true.

Proof. The condition "$\phi \leq \psi$" between two Boolean functions states that, for each set of values of the independent variables, the corresponding value of ϕ is less than or equal to that of ψ. Evidently, if the value of ϕ is zero, this condition is automatically satisfied no matter what the value of ψ is. Thus, "$\phi \leq \psi$" is equivalent to the requirement that whenever ϕ has the value 1, the value of ψ must also be 1. But the correspondence between truth value functions and Boolean functions enables us to phrase this requirement as: "Whenever f has a value which is true, the value for g must also be true." This is equivalent to "$f \to g$". ∎

By the method used in proving Theorem 5.1, any true sentence about the ordering relation in $\mathcal{B}_n$ can be used to obtain a true sentence about truth value functions. Some of the more important cases are given in the following theorem.

THEOREM 5.2 For any truth value functions f, g, and h, of the same n independent variables,

(a) $f \Rightarrow f$.

(b) $(f \to g) \wedge (g \to f) \Leftrightarrow (f \leftrightarrow g)$.

(c) $(f \to g) \wedge (g \to h) \Rightarrow (f \to h)$.

(d) $(f \leftrightarrow 1) \wedge (f \to g) \Rightarrow (g \leftrightarrow 1)$.

Proof. Parts (a), (b), and (c) are left as exercises (Problem 1). For part (d), let ϕ and ψ be the Boolean functions which correspond, respectively, to f and g. The sentence "$f \leftrightarrow 1$" means that all the values of the function f are true sentences; that is, the Boolean function ϕ is identically 1, so that $\phi = G$ (the greatest element in $\mathcal{B}_n$). Thus, condition (d) can be formulated in terms of ϕ and ψ as follows

$$\phi = G \text{ and } \phi \leq \psi \Rightarrow \psi = G.$$

This result follows immediately from Problem 13, Section 3-2. ∎

The result of Theorem 5.2d is a well-known rule of logic called *modus ponens*. Because of this rule, the truth value functions f such that $f \Leftrightarrow 1$ (the corresponding Boolean function is identically 1) are particularly important. These truth value functions are the tautologies.

PROBLEMS

1. Complete the proof of Theorem 5.2.

2. If the truth tables of f and g are given, how can you decide whether or not f implies g? How can you decide whether or not f is equivalent to g?

3. Find two truth value functions f and g, of the same independent variables, such that both of "$f \rightarrow g$" and "$g \rightarrow f$" are false. Are there two sentences a and b such that both of "$a \rightarrow b$" and "$b \rightarrow a$" are false?

4. Let $f_1, f_2, \ldots, f_n$ be truth value functions of the same independent variables.

   (a) Prove $(f_1 \rightarrow f_2) \wedge (f_2 \rightarrow f_3) \wedge \ldots \wedge (f_{n-1} \rightarrow f_n) \Rightarrow (f_1 \rightarrow f_n)$.

   (b) Prove that, for all integers i, j between 1 and n inclusive,

$$(f_1 \rightarrow f_2) \wedge (f_2 \rightarrow f_3) \wedge \ldots \wedge (f_{n-1} \rightarrow f_n) \wedge (f_n \rightarrow f_1) \Rightarrow (f_i \rightarrow f_j).$$

5. What conclusions are possible from the following information?

   (a) None of the unnoticed things, met with on a space trip, are Martians.

   (b) Things entered in the log, as met with on a space trip, are sure to be worth remembering.

   (c) I have never met anything worth remembering, while on a space trip.

   (d) Things met with on a space trip, that are noticed, are sure to be recorded in the log.

6. What conclusions are possible from the following information?

   (a) The only coeds in this class are sorority girls.

   (b) Every girl is suitable for a date, that loves to gaze at the moon.

   (c) When I detest a girl, I avoid her.

   (d) No girl is sufficiently beautiful, unless she will dance all night.

   (e) No sorority girls fail to be wealthy.

   (f) No girls ever take to me, except those which are in this class.

   (g) Miss X is not suitable for a date.

   (h) None but sufficiently beautiful girls are wealthy.

   (i) I detest girls that do not take to me.

   (j) Girls that will dance all night always love to gaze at the moon.

7. Which truth value functions correspond to the least element of the ordered set $\mathcal{B}_n$?

PROBLEMS

4

LATTICES

4-1 Introduction

We have studied the algebraic system composed of the set $\mathcal{B}_n$ with an ordering relation and have found that this system is convenient for expressing implications or equivalences between truth value functions of the same n independent variables. In this chapter, we shall see that the ordering relation in $\mathcal{B}_n$ can be used to define two operations on $\mathcal{B}_n$ and these operations will be of interest in discussing the logical connectives "$\wedge$" and "$\vee$" between truth value functions. As usual, we shall present the abstract situation first and then specialize to the particular ordered set $\mathcal{B}_n$. The basic definitions are given in Section 4-2, and further properties of the algebraic system are given in Section 4-3. In Section 4-4 the general theory is specialized to $\mathcal{B}_n$ and the application to logic is discussed.

[*Caution:* In the remainder of the text we shall be using the metalanguage much more freely than we have been. For example, we may write simply "$a \leq c$" instead of "'$a \leq c$' is true" as we have done in the first three chapters. The context will indicate whether a sentence is being asserted as true, or is merely being presented for consideration.]

4-2 Basic Definitions

In this section we shall study certain ordered sets as algebraic systems. We shall need some new terminology to state the definition.

Let S be an ordered set and let A be a subset of S. An *upper bound* of A is an element c of S such that, for each element a of A, $a \leq c$; a *lower bound* of A is an element d of S such that, for each element a of A, $d \leq a$. Of course, there may be several upper bounds for a particular set A, or there may be none at all. For the ordered set S pictured in Fig. 2.1a of Section 3-2, the set $A = \{e, d, g\}$ has the 4 elements a, b, c, d as upper bounds; the set $A' = \{a, f, c\}$ has no upper bounds. Evidently, if c is an upper bound for the set A and if $c \leq c'$, then c' is also an upper bound of A. This suggests that it might be convenient to choose the least of the upper bounds for A (if there is a least one) to represent all of the upper bounds.

DEFINITION 2.1 Let A be a subset of an ordered set S. The *supremum* of A, in symbols $\bigcup A$, or sup A, is an element of S such that

$$\begin{aligned} &\sup A \text{ is an upper bound of } A, \quad \text{and} \\ &\text{if } c \text{ is any upper bound of } A, \quad \text{then } \sup A \leq c. \end{aligned}$$

Similarly, the *infimum* of A, in symbols $\bigcap A$, or inf A, is an element of S such that

$$\begin{aligned} &\inf A \text{ is a lower bound of } A, \quad \text{and} \\ &\text{if } d \text{ is any lower bound of } A, \quad \text{then } d \leq \inf A. \end{aligned}$$

In case the set A is finite, say $A = \{a_1, a_2, \ldots, a_n\}$, the elements $\bigcup A$ and $\bigcap A$ are sometimes denoted by $a_1 \cup a_2 \cup \ldots \cup a_n$ and $a_1 \cap a_2 \cap \ldots \cap a_n$ respectively. Also, the supremum of a set A will be referred to as the supremum of the elements of the set; for example, we may speak of the sup of a and b instead of sup $\{a, b\}$. Similarly with the infimum.

Example 2.2 (a) For the ordered set S pictured in Fig. 2.1b of Section 3-2,

$$\begin{aligned} \sup \{c, d\} &= c \cup d = a, \\ \inf \{a, b, e\} &= a \cap b \cap e = b. \end{aligned}$$

(b) Since a subset of an ordered set may have no upper bounds, it certainly may fail to have a sup. Moreover, sup A may not exist even when A does have an upper bound. Let S be the set of non-zero real numbers with the usual ordering and let A be the subset composed of the negative numbers. Any positive number is an upper bound for A, but sup A does not exist since the set of all upper bounds of A does not have a least element.

DEFINITION 2.3 A *lattice* is an ordered set such that each pair of elements has both a sup and an inf.

We have already seen examples of some ordered sets which are lattices and some which are not. The ordered set of Fig. 2.1a in Section 3-2 fails to be a lattice (for instance, e and f have no inf); it is easy to check that the ordered set of Fig. 2.1b in Section 3-2 is a lattice. The ordered sets of Examples 2.3a and 2.3b in Section 3-2 are lattices.

Example 2.4 Let S be a set and let $\mathfrak{S}$ be the collection of all subsets of S, ordered by inclusion. Then $\mathfrak{S}$ is a lattice. In fact, if A and B are any two elements of $\mathfrak{S}$, the sup of A and B is just the union of the two sets; that is,

$A \cup B =$ the set of all objects which are elements of at least one of the two sets A and B.

Similarly, the inf of A and B is their intersection.

$A \cap B =$ the set of all objects which are elements of both of the sets A and B.

We shall use the symbols "$\smile$" and "$\frown$" for the set operations union and intersection, respectively. That is, we shall write "$A \smile B$" for the union of the sets A and B, and shall write "$A \frown B$" for the intersection of A and B.

The operations sup and inf are not really binary operations on an ordered set S since it is not necessary to have exactly a pair of elements in order to perform these operations. Indeed, as Definition 2.1 suggests, it may be possible to perform these operations with any subset of S, finite or infinite, or with the entire set S. Of course, there may be subsets of S which do not have a sup or an inf; that is, subsets such that the operations sup or inf cannot be performed (Problem 3). The characteristic property of a lattice is that each of the operations sup and inf can be performed with any pair of elements. Thus, in a lattice, it is frequently convenient to consider $\cup$ and $\cap$ as binary operations. The following theorem shows that these operations are defined for all non-empty finite sets. The situation with respect to infinite sets is presented in Problem 8.

THEOREM 2.5 In a lattice, any non-empty subset consisting of a finite number of elements has both a sup and an inf.

Proof. Problem 7. ∎

It is instructive to consider the binary operations "$\cup$" and "$\cap$" defined on a lattice S by setting

$$a \cup b = \cup\{a, b\} \quad \text{and} \quad a \cap b = \cap\{a, b\}.$$

The following theorem gives several interesting properties of these operations.

THEOREM 2.6 Let a, b, and c be elements of a lattice S. The binary operations "$\cup$" and "$\cap$" on S have the following properties:

Commutative Laws;

$a \cup b = b \cup a;$ $\qquad$ $a \cap b = b \cap a.$

Associative Laws;

$a \cup (b \cup c) = (a \cup b) \cup c;$ $\qquad$ $a \cap (b \cap c) = (a \cap b) \cap c.$

Idempotent Laws;

$a \cup a = a;$ $\qquad$ $a \cap a = a.$

Absorption Laws;

$a \cup (a \cap b) = a;$ $\qquad$ $a \cap (a \cup b) = a.$

Proof. We shall prove two of these laws; the proofs of the others are left as exercises (Problem 9).

Proof that $a \cup b = b \cup a$: By definition,

$$a \cup b = \sup\{a, b\} \quad \text{and} \quad b \cup a = \sup\{b, a\}.$$

but the sets $\{a, b\}$ and $\{b, a\}$ are identical since they have the same elements. Thus $a \cup b = b \cup a$.

Proof that $a \cup (a \cap b) = a$: Since the ordering relation is reflexive, we must have

$$a \leq a.$$

Also, since $a \cap b$ is one of the lower bounds for $\{a, b\}$, we have

$$a \cap b \leq a.$$

These two relations show that the element a is one of the upper bounds of the set $\{a, a \cap b\}$. Evidently, if c is any upper bound of $\{a, a \cap b\}$, then $a \leq c$. Thus, by Definition 2.1, sup $\{a, a \cap b\} = a$.

PROBLEMS

1. Prove that each subset of an ordered set has at most one sup and at most one inf.

2. In the ordered set of Fig. 2.1b in Section 3-2, find the sup and the inf of each subset consisting of either one or two elements.

3. Find an example of an ordered set S with a subset A, containing exactly two elements, such that A has an upper bound but no sup.

4. Let S be the set of all points (n, m) in the plane with positive integral coordinates. In which of the following cases is S an ordered set? In which cases is it a lattice? What is the geometric interpretation of the binary operations "$\cup$" and "$\cap$" in each of the lattices?

   (a) $(n, m) \leq (n', m') \Leftrightarrow n \leq n'$ and $m \leq m'$.

   (b) $(n, m) \leq (n', m') \Leftrightarrow n \leq m'$ and $m \leq n'$.

   (c) $(n, m) \leq (n', m') \Leftrightarrow n \leq n'$ and $m' \leq m$.

   (d) $(n, m) \leq (n', m') \Leftrightarrow n \leq m$ and $n' \leq m'$.

   (e) $(n, m) \leq (n', m') \Leftrightarrow n \leq m$.

   (f) $(n, m) \leq (n', m') \Leftrightarrow n$ is a factor of n' and m is a factor of m'.

   (g) $(n, m) \leq (n', m') \Leftrightarrow 2n + m \leq 2n' + m'$.

   (h) $(n, m) \leq (n', m') \Leftrightarrow 2n - m \leq 2n' - m'$.

5. Prove that each of the ordered sets of Examples 2.3a and 2.3b in Section 3-2 are lattices. What are the sup and inf of pairs of elements in these lattices?

6. Find an example of a collection $\mathcal{S}$ of sets which is ordered by inclusion but does not form a lattice.

7. Prove Theorem 2.5.

8. (a) Find an example of an infinite subset A of a lattice such that A has no upper bounds.

   (b) Find an example of an infinite subset A of a lattice such that A has an upper bound but no sup.

9. Complete the proof of Theorem 2.6.

10. Draw the graphical representation of the ordered set $S = \{1, 2, 3, \ldots, 10\}$ with the ordering relation defined by

$$a \leq b \Leftrightarrow a \text{ has fewer factors than } b, \text{ or } a = b.$$

Is this ordered set a lattice?

11. Draw the graphical representation of the ordered set $S = \{1, 2, 3, \ldots, 10\}$ with the ordering relation defined by

$a \leq b \Leftrightarrow a$ has fewer factors than b, or

a and b have the same number of factors and a is less than or equal to b.

Is this ordered set a lattice?

12. With the notation of Problem 14, Section 3-2, prove that $\mathbb{S}$, with the ordering relation of inclusion, is a lattice if and only if

$$(A \leq B \text{ and } A_1 \leq B_1) \Rightarrow (A \smile A_1 \leq B \smile B_1)$$

and

$$(A \leq B \text{ and } A_1 \leq B_1) \Rightarrow (A \frown A_1 \leq B \frown B_1).$$

4-3 Further Properties of Lattices

Theorem 2.6 lists several properties of the binary operations "$\cup$" and "$\cap$" in a lattice. Many other properties of these operations could have been mentioned, but the ones given are of particular importance; the following theorem shows that these properties actually characterize lattices.

THEOREM 3.1 If an algebraic system is composed of a set S and two binary operations "$*$" and "$\bigcirc$" on S such that, for all elements a, b, and c of S,

(i) $a * b = b * a$, $\quad a \bigcirc b = b \bigcirc a$;

(ii) $a * a = a$, $\quad a \bigcirc a = a$;

(iii) $a * (b * c) = (a * b) * c$, $\quad a \bigcirc (b \bigcirc c) = (a \bigcirc b) \bigcirc c$;

(iv) $a * (a \bigcirc b) = a$, $\quad a \bigcirc (a * b) = a$;

then there is a unique ordering relation in S which makes S a lattice and such that the given operations "$*$" and "$\bigcirc$" are, respectively, "$\cup$" and "$\cap$" in the lattice.

Proof. The proof consists of Lemmas 3.2 through 3.4 below.

LEMMA 3.2 With S, "$*$", and "$\bigcirc$" as in Theorem 3.1,

$$a * b = b \Leftrightarrow a \bigcirc b = a.$$

Proof. Suppose $a * b = b$. Then

$$a \circ b = a \circ (a * b) = a$$

by the second of conditions (iv) in Theorem 3.1. The proof of the converse is similar. ∎

LEMMA 3.3 With S, "$*$", and "$\circ$" as in Theorem 3.1, define "$\leq$" by

$$a \leq b \Leftrightarrow a * b = b.$$

Then "$\leq$" is an ordering relation on S; moreover, with this ordering relation, S is a lattice and the operations "$\cup$" and "$\cap$" in the lattice are, respectively, "$*$" and "$\circ$".

Proof. It is evident that "$\leq$" is a relation on S (for each pair of elements a, b of S, "$a \leq b$" is a sentence); to prove it is an ordering relation, we must show that it is reflexive, antisymmetric, and transitive.

The relation "$\leq$" is reflexive since, by the first of conditions (ii) in Theorem 3.1, $a * a = a$ for each element a in S. Thus $a \leq a \Leftrightarrow 1$.

The relation "$\leq$" is antisymmetric since, if both of "$a \leq b$" and "$b \leq a$" are true, then

$$a * b = b \quad \text{and} \quad b * a = a.$$

These equations, together with the first of conditions (i) in Theorem 3.1, imply $a = b$.

The relation "$\leq$" is transitive since, if $a * b = b$ and $b * c = c$, then

$$a * c = a * (b * c) = (a * b) * c = b * c = c.$$

This completes the proof that "$\leq$" is an ordering relation on S.

To complete the proof of Lemma 3.3 we must show that each pair of elements a, b of S has a sup and an inf and that

$$a \cup b = a * b \quad \text{and} \quad a \cap b = a \circ b.$$

First, notice that

$$a \leq a * b$$

because

$$a * (a * b) = (a * a) * b = a * b,$$

and that

$$b \leq a * b$$

because

$$b * (a * b) = a * (b * b) = a * b.$$

Thus $a * b$ is an upper bound for $\{a, b\}$. Let c be any upper bound for $\{a, b\}$; then

$$a * c = c \quad \text{and} \quad b * c = c,$$

so

$$(a * b) * c = a * (b * c) = a * c = c.$$

Hence $a * b \leq c$; this proves that $a * b = a \cup b$. The proof that $a \circ b = a \cap b$ is similar. ∎

LEMMA 3.4 With S, "$*$", and "$\circ$" as in Theorem 3.1, the relation "$\leq$" defined in Lemma 3.3 is the only ordering relation on S which makes S a lattice and such that the given operations "$*$" and "$\circ$" are, respectively, "$\cup$" and "$\cap$" in the lattice.

Proof. Let "$\prec$" be any ordering relation on S which makes S a lattice, and such that "$*$" and "$\circ$" are, respectively, "$\cup$" and "$\cap$". If $a \prec b$, then $a \cup b = b$. Therefore, since "$\cup$" is the same operation as "$*$", we must have $a * b = b$; thus, by the definition in Lemma 3.3, $a \leq b$. The proof is completed by noting that each of the steps above is reversible. ∎

This completes the proof of Theorem 3.1. ∎

So far, in Sections 4-2 and 4-3 we have considered only conditions which are satisfied by all lattices. There are several interesting conditions which are satisfied by some lattices, but not by others. The remainder of this section is devoted to such conditions.

DEFINITION 3.5 A lattice S is *distributive* iff, for all elements a, b, c of S,

$$a \cap (b \cup c) = (a \cap b) \cup (a \cap c).$$

Thus, a lattice is distributive iff the operation "$\cap$" distributes over the operation "$\cup$". This seemingly unsymmetric treatment of the two lattice operations is misleading; the condition is actually symmetric in the two operations, as the following theorem shows.

THEOREM 3.6 A lattice S is distributive iff, for all elements a, b, c, of S,

$$a \cup (b \cap c) = (a \cup b) \cap (a \cup c).$$

Proof. Suppose S is distributive; then

$$(a \cup b) \cap (a \cup c) = [(a \cup b) \cap a] \cup [(a \cup b) \cap c].$$

By the absorption law and another use of the distributive property,

$$[(a \cup b) \cap a] \cup [(a \cup b) \cap c] = a \cup [(a \cap c) \cup (b \cap c)].$$

The associative law, and another use of the absorption law, give

$$a \cup [(a \cap c) \cup (b \cap c)] = a \cup (b \cap c).$$

Thus, if S is distributive, then

$$a \cup (b \cap c) = (a \cup b) \cap (a \cup c).$$

The proof of the converse is left as an exercise (Problem 3). ∎

Example 3.7 The lattice of all subsets of a given set S, ordered by inclusion, is a distributive lattice.

Proof. If A, B, and C are subsets of S, the set

$$A \frown (B \smile C)$$

is composed of all elements which are in A and also in at least one of B or C. The set

$$(A \frown B) \smile (A \frown C)$$

is composed of all elements which are either in both of A and B or in both of A and C. It is easy to see that these two conditions on the elements of S are equivalent; thus

$$A \frown (B \smile C) = (A \frown B) \smile (A \frown C).$$ ∎

The construction of an example of a non-distributive lattice is left as an exercise (Problem 4).

We have seen (Problem 9 of Section 3-2) that an ordered set may or may not contain greatest and least elements. The same is true of a lattice. The real numbers with the usual ordering form a lattice with neither a greatest nor a least element; the real numbers between zero and one inclusive form a lattice with both a greatest and a least element. Further examples appear in the problems.

As we have seen, the collection $\mathfrak{S}$ of all subsets of a given set S forms a lattice when inclusion is the ordering relation. The greatest element G of this lattice is S itself, and the least element L is the empty set $\emptyset$. The familiar set operations of union and intersection are the operations sup and inf in the lattice. But there is another set operation, complementation, which we have not yet had occasion to use. The complement of a subset A of S is defined to be the collection of all elements of S which are not elements of A. Thus, complementation is a unary operation (performed on a single set). If A' is the complement of A, it is easy to see that

$$A \smile A' = S$$

and

$$A \frown A' = \emptyset.$$

This familar set operation suggests the following definition.

DEFINITION 3.8 In a lattice S with greatest element G and least element L, a *complement* of the element a of S is an element b of S such that

$$a \cup b = G \quad \text{and} \quad a \cap b = L.$$

A *complemented lattice* is a lattice in which there is a greatest element and a least element and in which each element has at least one complement.

PROBLEMS

1. Which of the following ordered sets are lattices? Which of the lattices are distributive? Which have greatest or least elements? Which are complemented?

(a) The set $\mathfrak{S}$ of all finite subsets of the real numbers, ordered by inclusion.

(b) The set $\mathfrak{S}$ of all infinite subsets of the real numbers, ordered by inclusion.

(c) The set $\mathfrak{S}$ of all complements of finite subsets of the real numbers, ordered by inclusion.

(d) The set $\mathfrak{S}$ of all subsets of the real numbers which have an even number of elements, ordered by inclusion.

(e) The set $\mathfrak{S}$ of all subsets of the real numbers which have the set $\{0, 1\}$ as a subset, ordered by inclusion.

2. Which of the lattices of Problem 4 in Section 4-2 are distributive? Which have a greatest or least element? Which are complemented?

3. Complete the proof of Theorem 3.6.

4. Find two distributive lattices and two which are not distributive. (*Hint:* There are two non-distributive lattices, each with five elements, which are not isomorphic.)

5. (a) Let S be a lattice and let b and c be two elements of S such that for each element a of S,

$$a \cup b = b \quad \text{and} \quad a \cup c = a.$$

Prove that $b = G$ and $c = L$.

(b) Similarly, characterize the elements G and L of a lattice in terms of the operation "$\cap$".

6. Find two complemented lattices and two which are not complemented.

7. Find a lattice that contains an element a having two different complements. Is it possible for every element in a lattice to have more than one complement?

8. Let S be ordinary three-dimensional space, and let the elements of $\mathcal{S}$ be the empty set, the sets consisting of the single points of S, the lines in S, the planes in S, and the set S itself. Prove that $\mathcal{S}$, with the ordering relation of inclusion, is a non-distributive, complemented lattice.

9. With the notation of Problem 14, Section 3-2, show that

$$(A \leq B \rightarrow B' \leq A') \Leftrightarrow (X \text{ is in } \mathcal{S} \rightarrow X' \text{ is in } \mathcal{S}).$$

(The operation "$'$" is the operation of point-set complementation.)

#10. Prove that, in a distributive lattice,

$$B \cap (A_1 \cup A_2 \cup \cdots \cup A_n) = (B \cap A_1) \cup (B \cap A_2) \cup \cdots \cup (B \cap A_n).$$

4-4 The Lattice $\mathcal{B}_n$ and Applications to Logic

An ordering relation was defined on the set $\mathcal{B}_n$ in Section 3-4 and some relationships between the ordered set $\mathcal{B}_n$ and logic were discussed in Section 3-5. In this section we shall see that the ordered set $\mathcal{B}_n$ is a lattice and that the lattice operations in $\mathcal{B}_n$ have interesting analogs in logic.

The concepts of the maximum and minimum of two real-valued functions should be familiar, but, for completeness, we recall the definitions for the special case in which we shall be interested. If ϕ and ψ are two Boolean functions of the same independent variables $\xi_1, \xi_2, \ldots, \xi_n$, then max $\{\phi, \psi\}$ is a Boolean function of these independent variables and, for each set $\alpha_1, \alpha_2, \ldots, \alpha_n$ of values of these independent variables, the corresponding value of the function,

$$\max \{\phi, \psi\} (\alpha_1, \alpha_2, \ldots, \alpha_n),$$

is the maximum of the two numbers

$$\phi(\alpha_1, \alpha_2, \ldots, \alpha_n)$$

and

$$\psi(\alpha_1, \alpha_2, \ldots, \alpha_n).$$

Similarly, min $\{\phi, \psi\}$ is a Boolean function whose value,

$$\min \{\phi, \psi\} (\alpha_1, \alpha_2, \ldots, \alpha_n),$$

is the minimum of the two numbers

$$\phi(\alpha_1, \alpha_2, \ldots, \alpha_n)$$

and

$$\psi(\alpha_1, \alpha_2, \ldots, \alpha_n).$$

THEOREM 4.1 The ordered set $\mathcal{B}_n$ is a complemented distributive lattice.

Proof. Let ϕ and ψ be two Boolean functions of n independent variables. Clearly

$$\phi \leq \max \{\phi, \psi\}$$

and

$$\psi \leq \max \{\phi, \psi\}.$$

Also, if Ω is an element of $\mathcal{B}_n$ such that

$$\phi \leq \Omega \quad \text{and} \quad \psi \leq \Omega,$$

then

$$\max \{\phi, \psi\} \leq \Omega.$$

Thus

$$\phi \cup \psi = \max \{\phi, \psi\}.$$

Similarly,

$$\phi \cap \psi = \min \{\phi, \psi\},$$

so $\mathcal{B}_n$ is a lattice.

To prove that $\mathcal{B}_n$ is distributive, let ϕ, ψ, and Ω be three elements of $\mathcal{B}_n$. Then

$$\phi \cap (\psi \cup \Omega) = \min \{\phi, \max \{\psi, \Omega\} \}.$$

Thus, the value of the function $\phi \cap (\psi \cup \Omega)$ is 1 for a particular set of values of the independent variables iff

(A) The function ϕ has the value 1 for these values, and at least one of the functions ψ and Ω has the value 1 for these values.

On the other hand,

$$(\phi \cap \psi) \cup (\phi \cap \Omega) = \max \{\min \{\phi, \psi\}, \min \{\phi, \Omega\} \}.$$

Thus, the value of the function $(\phi \cap \psi) \cup (\phi \cap \Omega)$ is 1 for a particular set of values of the independent variables iff

(B) Both of ϕ and ψ have the value 1 for these values, or both of ϕ and Ω have the value 1 for these values.

Since the conditions (**A**) and (**B**) are equivalent,

$$\phi \cap (\psi \cup \Omega) = (\phi \cap \psi) \cup (\phi \cap \Omega),$$

so the lattice $\mathcal{B}_n$ is distributive.

The proof that $\mathcal{B}_n$ is complemented is left as an exercise (Problem 1). ∎

As we have seen in Chapter 3, the correspondence between truth value functions and Boolean functions is useful in certain logical considerations. Since the ordered set $\mathcal{B}_n$ has now been shown to have further algebraic structure (as a lattice), it is natural to enquire what operations on truth value functions correspond to the lattice operations in $\mathcal{B}_n$. The result is given in Theorem 4.3 below, but a definition is needed first.

DEFINITION 4.2 If f and g are truth value functions of the same independent variables $x_1, x_2, \ldots, x_n$, then $f \vee g, f \wedge g$, and $\sim f$ are the truth value functions defined as follows:

$$(f \vee g)(x_1, x_2, \ldots, x_n) = f(x_1, x_2, \ldots, x_n) \vee g(x_1, x_2, \ldots, x_n)$$

$$(f \wedge g)(x_1, x_2, \ldots, x_n) = f(x_1, x_2, \ldots, x_n) \wedge g(x_1, x_2, \ldots, x_n)$$

$$(\sim f)(x_1, x_2, \ldots, x_n) = \sim f(x_1, x_2, \ldots, x_n).$$

THEOREM 4.3 If the truth value functions f, g (with the same independent variables) correspond, respectively, to the Boolean functions ϕ, ψ, then

$$f \vee g \quad \text{corresponds to} \quad \phi \cup \psi$$

and

$$f \wedge g \quad \text{corresponds to} \quad \phi \cap \psi.$$

Proof. The Boolean function which corresponds to $f \vee g$ has the value 1 iff at least one of f and g has a value which is true; that is, iff at least one of ϕ and ψ has the value 1. But this condition characterizes the Boolean function $\phi \cup \psi$. Similarly, the Boolean function which corresponds to $f \wedge g$ has the value 1 iff both of the functions f and g have values which are true; that is, iff both of ϕ and ψ have the value 1. This condition characterizes the Boolean function $\phi \cap \psi$. ∎

The correspondence between the operations "$\vee$" and "$\wedge$" for truth value functions and the operations "$\cup$" and "$\cap$" for Boolean functions, together with the relationship between implication and the ordering in $\mathcal{B}_n$, makes it possible to find several basic theorems of logic. Some of these familiar results are collected in the following theorem. Of course they could be obtained directly, without using Boolean functions.

THEOREM 4.4 If f and g are truth value functions of the same n independent variables, then

(a) $f \Rightarrow f \vee g$.

(b) $g \Rightarrow f \vee g$.

(c) $f \wedge g \Rightarrow f$.

(d) $f \wedge g \Rightarrow g$.

(e) $f \wedge g \Rightarrow f \vee g$.

(f) $(f \wedge g \leftrightarrow 1) \Leftrightarrow (f \leftrightarrow 1) \wedge (g \leftrightarrow 1)$.

(g) $(f \vee g \leftrightarrow 0) \Leftrightarrow (f \leftrightarrow 0) \wedge (g \leftrightarrow 0)$.

Proof. Let ϕ and ψ be the Boolean functions which correspond to f and g respectively. The implication (a) is immediate from the order relation

$$\phi \leq \phi \cup \psi.$$

Parts (b) through (e) are similar. To prove (f), notice that $f \wedge g \leftrightarrow 1$ is equivalent to

The Boolean function $\phi \cap \psi$ *is identically* 1.

This condition is equivalent to

Each of ϕ *and* ψ *is identically* 1.

This, in turn, is equivalent to

$$(f \leftrightarrow 1) \wedge (g \leftrightarrow 1).$$

The proof of (g) is similar. ∎

PROBLEMS

1. The following results complete the proof of Theorem 4.1.

(**a**) What are the greatest and least elements in the lattice $\mathcal{B}_n$?

(b) Prove that every element of $\mathcal{B}_n$ has a complement.

(c) Does every element of $\mathcal{B}_n$ have a unique complement? Can you express the complement of ϕ conveniently in terms of ϕ?

2. Complete the proof of Theorem 4.4 by supplying proofs for the parts not treated in the text.

3. Let f and g be truth value functions of the same independent variables. Which of the following sentences are true?

(a) $(f \vee g \leftrightarrow 1) \leftrightarrow (f \leftrightarrow 1) \vee (g \leftrightarrow 1)$.

(b) $(f \vee g \leftrightarrow 1) \leftrightarrow (f \leftrightarrow 1) \wedge (g \leftrightarrow 1)$.

(c) $(f \wedge g \leftrightarrow 1) \leftrightarrow (f \leftrightarrow 1) \vee (g \leftrightarrow 1)$.

(d) $(f \wedge g \leftrightarrow 1) \leftrightarrow (f \leftrightarrow 1) \wedge (g \leftrightarrow 1)$.

4. Repeat Problem 3 with the symbol "1" replaced by "0".

5. Let f and g be truth value functions of the same n independent variables, and let $a_1, a_2, \ldots, a_n$ be sentences in the respective ranges of these variables. Which of the following statements are true?

(a) $[(f \vee g)(a_1, a_2, \ldots, a_n) \leftrightarrow 1]$

$\leftrightarrow [f(a_1, a_2, \ldots, a_n) \leftrightarrow 1] \vee [g(a_1, a_2, \ldots, a_n) \leftrightarrow 1]$.

(b) $[(f \vee g)(a_1, a_2, \ldots, a_n) \leftrightarrow 1]$

$\leftrightarrow [f(a_1, a_2, \ldots, a_n) \leftrightarrow 1] \wedge [g(a_1, a_2, \ldots, a_n) \leftrightarrow 1]$.

(c) $[(f \wedge g)(a_1, a_2, \ldots, a_n) \leftrightarrow 1]$

$\leftrightarrow [f(a_1, a_2, \ldots, a_n) \leftrightarrow 1] \vee [g(a_1, a_2, \ldots, a_n) \leftrightarrow 1]$.

(d) $[(f \wedge g)(a_1, a_2, \ldots, a_n) \leftrightarrow 1]$

$\leftrightarrow [f(a_1, a_2, \ldots, a_n) \leftrightarrow 1] \wedge [g(a_1, a_2, \ldots, a_n) \leftrightarrow 1]$.

6. Repeat Problem 5 with the symbol "1" replaced by "0".

5

BOOLEAN ALGEBRAS

5-1 Introduction

A Boolean algebra is defined in Section 5-2 and some of the elementary properties of this type of algebraic system are developed. In Section 5-3 an important representation theorem is presented; this theorem shows that every finite Boolean algebra can be represented by the collection of all subsets of some set. The collection $\mathcal{B}_n$ of all Boolean functions of n independent variables is a Boolean algebra. This particular Boolean algebra, and some of the applications to logic, are discussed in Section 5-4.

5-2 Basic Definition and Properties

We have studied ordered sets and have defined a lattice as a special type of ordered set. A Boolean algebra is a special type of lattice.

DEFINITION 2.1 A *Boolean algebra* is a complemented, distributive lattice.

Thus a Boolean algebra is an ordered set in which each pair of elements has both a sup and an inf, each of the binary operations "$\cup$" and "$\cap$" distributes over the other, there is a greatest element and a least element, and each element has at least one complement. Note that a Boolean algebra must be non-empty — in particular, it must contain a greatest element.

THEOREM 2.2 Each element in a Boolean algebra has precisely one complement.

NOTATION: We shall use "$\bar{A}$" to denote the complement of the element A in a Boolean algebra.

Proof. Suppose that B and C are each complements of the element A in a Boolean algebra $\mathcal{B}$, then

$$B = B \cap G = B \cap (A \cup C) = (B \cap A) \cup (B \cap C)$$
$$= L \cup (C \cap B) = C \cap B.$$

and

$$C = C \cap G = C \cap (A \cup B) = (C \cap A) \cup (C \cap B)$$
$$= L \cup (C \cap B) = C \cap B.$$

Thus $B = C$. ∎

Example 2.3 According to Theorem 4.1 of Section 4-4, the set $\mathcal{B}_n$ of all Boolean functions of n independent variables is a Boolean algebra. For each element ϕ of $\mathcal{B}_n$, its complement is given by

$$\bar{\phi} = 1 - \phi.$$

If the truth value function f corresponds to the Boolean function ϕ, then the truth value function $\sim f$ corresponds to the Boolean function $\bar{\phi}$.

Example 2.4 Let S be any set, and let $\mathfrak{S}$ be the collection of all subsets of S, ordered by inclusion. Then $\mathfrak{S}$ is a Boolean algebra.

The binary operation "$\cup$" is union: $A \cup B = A \smile B$.

The binary operation "$\cap$" is intersection:

$$A \cap B = A \frown B.$$

The least element is the empty set: $L = \emptyset$.

The greatest element is the set S: $G = S$.

The unary operation "$\bar{\ }$" is the set operation complementation: $\bar{A} = A'$.

The Boolean algebra $\mathfrak{S}$ of Example 2.4 shows that the Boolean operation of complement is connected in some way to the set operation of complement. The next theorem strengthens this connection by showing that, in any Boolean algebra, the operation "$\bar{\ }$" has properties analogous to those of "$'$".

THEOREM 2.5 In any Boolean algebra, the operation "$\bar{\ }$" satisfies the following conditions.

(a) $\bar{\bar{A}} = A$.

(b) $\overline{A \cup B} = \bar{A} \cap \bar{B}$.

(c) $\overline{A \cap B} = \bar{A} \cup \bar{B}$.

De Morgan's Laws.

(b′) $A \cup B = \overline{\bar{A} \cap \bar{B}}$.

(c′) $A \cap B = \overline{\bar{A} \cup \bar{B}}$.

(d) $\bar{L} = G$.

(e) $\bar{G} = L$.

Proof. We shall prove only part (b); the remainder of the proof is left as an exercise (Problem 1). Since complements are unique in a Boolean algebra, if we find any element X which satisfies

$$(A \cup B) \cup X = G$$

and

$$(A \cup B) \cap X = L,$$

then X must be the complement of $A \cup B$. It remains merely to try $X = \bar{A} \cap \bar{B}$. We find

$$\begin{aligned}(A \cup B) \cup (\bar{A} \cap \bar{B}) &= [(A \cup B) \cup \bar{A}] \cap [(A \cup B) \cup \bar{B}] \\ &= G \cap G = G\end{aligned}$$

and

$$\begin{aligned}(A \cup B) \cap (\bar{A} \cap \bar{B}) &= [A \cap (\bar{A} \cap \bar{B})] \cup [B \cap (\bar{A} \cap \bar{B})] \\ &= L \cup L = L. \blacksquare\end{aligned}$$

A characterization of a lattice was given in Theorem 3.1, Section 4-3, in terms of properties of two binary operations "$*$" and "$\circ$" which were proved to be the operations "$\cup$" and "$\cap$" in the lattice, respectively. The following theorem gives a similar characterization of a Boolean algebra.

THEOREM 2.6 Let an algebraic system be composed of a set $\mathfrak{S}$ containing (at least) two special elements P and Q, and of two binary operations "$*$" and "$\circ$" on $\mathfrak{S}$ such that, for any elements A, B, and C of $\mathfrak{S}$,

(a) $A * B = B * A$. $\qquad A \circ B = B \circ A$.

(b) $A * A = A$. $\qquad A \circ A = A$.

(c) $A * (B * C) = (A * B) * C$. $\qquad A \circ (B \circ C) = (A \circ B) \circ C$.

(d) $A * (A \circ B) = A$. $\qquad A \circ (A * B) = A$.

(e) $A \circ (B * C) = (A \circ B) * (A \circ C)$.

(f) For each element A of $\mathfrak{S}$ there is an element $\bar{A}$ of $\mathfrak{S}$ such that

$$A * \bar{A} = P \quad \text{and} \quad A \circ \bar{A} = Q.$$

Then there is a unique ordering relation in $\mathfrak{S}$ which makes $\mathfrak{S}$ a Boolean algebra and such that the given operations "$*$" and "$\circ$" are, respectively, "$\cup$" and "$\cap$" in the Boolean algebra. Moreover, P and Q are G and L, respectively, and $\bar{A}$ is the complement of A.

Proof. By Theorem 3.1, Section 4-3, the conditions (a) through (d) in Theorem 2.6 imply that there is a unique ordering in $\mathfrak{S}$ which makes $\mathfrak{S}$ a lattice, and such that "$*$" and "$\circ$" are, respectively, "$\cup$" and "$\cap$". Condition (e) states that this lattice is distributive.

For any element A of $\mathfrak{S}$,

$$A * P = A * (A * \bar{A}) = (A * A) * \bar{A} = A * \bar{A} = P$$

and

$$A \circ Q = A \circ (A \circ \bar{A}) = (A \circ A) \circ \bar{A} = A \circ \bar{A} = Q,$$

thus P and Q are, respectively, the greatest and least elements of the lattice. Condition (f) now states that the lattice is complemented and $\bar{A}$ is the complement of A. ∎

The number of elements in a finite set $\mathfrak{S}$ must satisfy a condition if $\mathfrak{S}$ can be made into a Boolean algebra. A necessary and sufficient condition will be found in the discussion of the next section, but the next theorem gives a simple necessary condition.

THEOREM 2.7 Every finite Boolean algebra $\mathfrak{B}$ (with one trivial exception) has an even number of elements.

Proof. See Problem 7 for the exception; it will be ignored in this proof. Form a decomposition of $\mathfrak{B}$ into two sets $\mathfrak{D}$ and $\mathfrak{E}$ as follows. The least element, L, is put in $\mathfrak{D}$; the greatest element, G, is put in $\mathfrak{E}$. If there is an element of $\mathfrak{B}$ which has not yet been put in either $\mathfrak{D}$ or $\mathfrak{E}$, let A be such an element. Now $\bar{A}$ must be different from A. (If $\bar{A} = A$, then $L = A \cap \bar{A} = A \cap A = A$ is already in $\mathfrak{D}$.) Also, $\bar{A}$ cannot already be in $\mathfrak{D}$ or in $\mathfrak{E}$. (If $\bar{A}$ is in $\mathfrak{D}$, for example, then $\bar{\bar{A}} = A$ is in $\mathfrak{E}$.) We put A in $\mathfrak{D}$ and $\bar{A}$ in $\mathfrak{E}$. Continuing recursively, we place elements in $\mathfrak{D}$ and their respective complements in $\mathfrak{E}$ until $\mathfrak{B}$ is exhausted and the sets $\mathfrak{D}$, $\mathfrak{E}$ form a decomposition of $\mathfrak{B}$. Since $\mathfrak{D}$ and $\mathfrak{E}$ have the same number of elements, the number of elements in $\mathfrak{B}$ must be even. ▮

PROBLEMS

1. Complete the proof of Theorem 2.5.

2. Is the set $\{0, 1, 2\}$ ordered by "less than or equal to" a Boolean algebra?

3. Prove that, in any Boolean algebra, the operation "$\cup$" can be expressed in terms of "$\cap$" and "$^{-}$"; prove also that "$\cap$" can be expressed in terms of "$\cup$" and "$^{-}$".

4. Use the result of Problem 3 to write the expression

$$[(A \cup B) \cap (C \cup D)] \cup (E \cap F)$$

(a) Without the operation "$\cup$".

(b) Without the operation "$\cap$".

5. (a) Given a set $\mathfrak{S}$ with a unary operation "$*$" and a binary operation "$\sim$", find conditions on "$*$" and "$\sim$" which are necessary and sufficient for the existence of an ordering in $\mathfrak{S}$ which makes $\mathfrak{S}$ a Boolean algebra in which "$*$" and "$\sim$" are "$\cup$" and "$-$" respectively.

(b) Similarly, characterize Boolean algebras in terms of the operations "$\cap$" and "$-$".

#6. Prove that each of the following conditions, on a particular pair A, B of elements of a Boolean algebra $\mathfrak{B}$, is necessary and sufficient for all of them.

(a) $A \leq B$.

(b) $\bar{B} \leq \bar{A}$.

(c) $A \cap \bar{B} \leq \bar{A}$.

(d) There is an element C of $\mathcal{B}$ such that $A \cap \bar{B} \leq C \cap \bar{C}$.

(e) $A \cap \bar{B} \leq L$.

(f) $\bar{A} \cup B \geq G$.

(*Hint:* Use Lemma 3.2, Section 4-3 and the definitions of sup and inf.)

7. Discuss Theorem 2.7 and its proof. What is the trivial exception? Why does the proof not apply to this exceptional case? Which step(s) in the proof are not valid in this case?

8. (a) Prove that, for every integer $n \geq 0$, there is a Boolean algebra having 2^n elements.

*(b) Prove that there is no Boolean algebra with 6 elements.

9. With the notation of Problem 14, Section 3-2, prove that $\mathfrak{S}$, with the ordering relation of inclusion, is a Boolean algebra if and only if, for any elements A, B, C, D of $\mathfrak{S}$,

$$A \leq B \Rightarrow B' \leq A'$$

and

$$(A \leq B) \wedge (C \leq D) \Rightarrow (A \smile C) \leq (B \smile D).$$

(The operation "$'$" is point-set complementation.)

5-3 Representation in Terms of Subsets of a Set

We have seen in Example 2.4 that the collection of all subsets of a given set S forms a Boolean algebra (when ordered by inclusion). Let us temporarily call such Boolean algebras set-algebras. Note that a set-algebra is not just a Boolean algebra whose elements are sets. It must be composed of *all* of the subsets of some set, and the ordering must be inclusion. The main result of this section is that every finite Boolean algebra is isomorphic to a set-algebra. Actually any Boolean algebra (finite or not) is isomorphic to a Boolean algebra of sets with the ordering relation being inclusion, and it can always be arranged that each of these sets is a subset of some particular set S. (The proof is beyond the scope of this text; see reference 27.) However, it may be that only some of the subsets of S are used; there are infinite Boolean algebras which are not isomorphic to the collection of all subsets of any set S (Problem 5).

THEOREM 3.1 Every finite Boolean algebra is isomorphic to the Boolean algebra of all the subsets of some set.

Proof. The proof consists of Lemmas 3.3 through 3.7 below. We shall need a definition and some notation.

DEFINITION 3.2 An element A of a Boolean algebra is an *atom* iff $X \leq A$ has exactly 2 solutions.

This definition can be rephrased as follows: A is an atom if and only if A is different from L and the two elements A and L are the only solutions to $X \leq A$.

The following notation will be used in the proof of Theorem 3.1.

$\mathcal{B}$: A (given) finite Boolean algebra.

$\mathcal{L}_B$: The set of all elements X of $\mathcal{B}$ such that $X \leq B$.

$\mathcal{A}$: The set of all atoms of $\mathcal{B}$.

$\mathcal{A}_B$: The set of all atoms A such that $A \leq B$. This is $\mathcal{A} \frown \mathcal{L}_B$.

LEMMA 3.3 For any element B of $\mathcal{B}$, the set $\mathcal{L}_B$ (with the ordering in $\mathcal{B}$) is a Boolean algebra. Moreover, if "$*$", "$\circ$", and "$\sim$" are, respectively, sup, inf, and complement in $\mathcal{L}_B$, then, for any elements C and D of $\mathcal{L}_B$

$$C * D = C \cup D$$

$$C \circ D = C \cap D$$

$$\tilde{C} = \bar{C} \cap B.$$

(We use "$\cup$", "$\cap$", and "$^{-}$" for sup, inf, and complement, respectively, in $\mathcal{B}$.)

Proof. For any elements C and D of $\mathcal{L}_B$,

$$C \leq B \quad \text{and} \quad D \leq B;$$

but then

$$C \cup D \leq B.$$

Thus $C \cup D$ is an element of $\mathcal{L}_B$. Since $C \cup D$ is the least of all the upper bounds of the set $\{C, D\}$ in $\mathcal{B}$, it certainly is the least of all the upper bounds of $\{C, D\}$ in $\mathcal{L}_B$. Hence

$$C * D = C \cup D.$$

The proof that $C \circ D = C \cap D$ is similar. Clearly the distributive law is satisfied in $\mathcal{L}_B$ since it holds in $\mathcal{B}$.

As to $\tilde{C}$, first note that, in $\mathfrak{L}_B$, the least and greatest elements are, respectively, L and B. Moreover, for any element C of $\mathfrak{L}_B$,

$$C \cup (\bar{C} \cap B) = (C \cup \bar{C}) \cap (C \cup B) = G \cap B = B$$

and

$$C \cap (\bar{C} \cap B) = L.$$

Since $\bar{C} \cap B$ is an element of $\mathfrak{L}_B$, these two equations show that

$$\tilde{C} = \bar{C} \cap B. \blacksquare$$

LEMMA 3.4 For any element $B \neq L$, of $\mathfrak{B}$,

$$\mathfrak{A} \frown \mathfrak{L}_B \neq \emptyset.$$

Proof. We must show that there is at least one atom A such that $A \leq B$. If B is an atom, the result is evident. If B is not an atom, there are at least three solutions for $X \leq B$. Let C be any solution of this inequality different from L and B. If C is an atom, the result is evident; if not, let D be a solution of $X \leq C$ which is different from L and C. Proceeding in this way, since $\mathfrak{B}$ is finite, we must arrive at an atom after a finite number of steps. $\blacksquare$

LEMMA 3.5 For any element B of $\mathfrak{B}$,

$$\sup \mathfrak{A}_B = B.$$

Proof. The case $B = L$ is left as an exercise (Problem 2). We assume that $B \neq L$. The proof is by contradiction. Since $\mathfrak{A}_B$ is a non-empty finite set, it has a supremum; set $\sup \mathfrak{A}_B = C$ and assume $C \neq B$. Clearly $C \leq B$ since B is one of the upper bounds of $\mathfrak{A}_B$. Thus C is an element of $\mathfrak{L}_B$. By Lemma 3.3, there is an element $\tilde{C}$ of $\mathfrak{L}_B$ such that

$$C \cup \tilde{C} = B \quad \text{and} \quad C \cap \tilde{C} = L.$$

Clearly $\tilde{C} \neq L$, since $C \cup L = C \neq B$. Thus, by Lemma 3.4, there is an atom $A \leq \tilde{C}$. Since $\tilde{C} \leq B$, A is an element of $\mathfrak{A}_B$ and it follows that $A \leq C$. Thus

$$C \cap \tilde{C} \geq A$$

in contradiction to the equation $C \cap \tilde{C} = L$. $\blacksquare$

LEMMA 3.6 Let $\mathfrak{A}'$ be any subset of $\mathfrak{A}$ and let $\sup \mathfrak{A}' = B$. Then $\mathfrak{A}_B = \mathfrak{A}'$.

Proof. The proof is by contradiction. Suppose

$$\mathcal{A}' = \{A_1, A_2, \ldots, A_n\}$$

is a set of atoms for which the lemma is false. Since

$$B = A_1 \cup A_2 \cup \cdots \cup A_n,$$

it follows that

$$A_i \leq B \ (i = 1, 2, \cdots, n).$$

Thus

$$\mathcal{A}' \subset \mathcal{A}_B.$$

Since, by assumption, $\mathcal{A}' \neq \mathcal{A}_B$, there must be an atom A_{n+1} in $\mathcal{A}_B$ but not in $\mathcal{A}'$. But this contradicts the distributive law since

$$A_{n+1} \cap (A_1 \cup A_2 \cup \cdots \cup A_n) = A_{n+1} \cap B = A_{n+1}$$

whereas, by Problem 3,

$$(A_{n+1} \cap A_1) \cup (A_{n+1} \cap A_2) \cup \cdots \cup (A_{n+1} \cap A_n)$$
$$= L \cup L \cup \cdots \cup L = L. \blacksquare$$

LEMMA 3.7 $\mathcal{B}$ is isomorphic to the Boolean algebra of all subsets of $\mathcal{A}$.

Proof. Let each element B of $\mathcal{B}$ correspond to the subset $\mathcal{A}_B$ of $\mathcal{A}$. We shall prove that this correspondence is an isomorphism.

(i) The correspondence is one-to-one from $\mathcal{B}$ onto the collection of all subsets of $\mathcal{A}$. It is evident that each element B of $\mathcal{B}$ determines exactly one subset $\mathcal{A}_B$ of $\mathcal{A}$. Let B and C be two distinct elements of $\mathcal{B}$. By Lemma 3.5,

$$B = \sup \mathcal{A}_B \quad \text{and} \quad C = \sup \mathcal{A}_C.$$

Thus $\mathcal{A}_B \neq \mathcal{A}_C$; distinct elements of $\mathcal{B}$ correspond to different subsets of $\mathcal{A}$. Finally, let $\mathcal{A}'$ be any subset of $\mathcal{A}$. By Lemma 3.6, $B = \sup \mathcal{A}'$ is an element of $\mathcal{B}$ which corresponds to $\mathcal{A}'$.

(ii) The correspondence preserves the ordering relation; that is, $B \leq C$ in $\mathcal{B}$ if and only if $\mathcal{A}_B \subset \mathcal{A}_C$. If $B \leq C$, then certainly for any atom A such that $A \leq B$, it must also be true that $A \leq C$. Thus

$$\mathcal{A}_B \subset \mathcal{A}_C.$$

Conversely, since

$$B = \sup \mathcal{A}_B \quad \text{and} \quad C = \sup \mathcal{A}_C,$$

if $\mathcal{A}_B \subset \mathcal{A}_C$, then $B \leq C$.

(iii) Since each of the special elements and operations in a Boolean algebra can be characterized in terms of the order relation, paragraphs (i) and (ii) above show that the correspondence under consideration is an isomorphism. ∎

This completes the proof of Theorem 3.1. ∎

COROLLARY 3.8 The number of elements in a finite Boolean algebra is a power of 2.

Proof. Problem 6. ∎

PROBLEMS

1. The proof of Lemma 3.4 is based on mathematical induction, but the inductive step is glossed over as the proof is presented in the text. Point out where this occurs and give the proof in the usual format for mathematical induction.
2. Prove Lemma 3.5 for the special case $B = L$.
3. (a) If A_i and A_j are two distinct atoms in a Boolean algebra $\mathcal{B}$, prove that $A_i \cap A_j = L$. (This result was used in the proof of Lemma 3.6.)

   (b) Discuss the proof of Lemma 3.6 for the special cases $n = 1$. Is the proof in the text valid for this case? Is the result correct?
4. Consider the special case $n = 0$ in the proof of Lemma 3.6. Is the proof in the text valid for this case? If not, give an alternate proof.
5. Give an example of a Boolean algebra which is not isomorphic to a Boolean algebra formed by the collection of all subsets of any set with inclusion as the ordering relation. (*Hint:* Use some of the infinite sequences of Boolean constants.)
6. Prove Corollary 3.8.
7. With reference to part (iii) of the proof of Lemma 3.7, prove that, if $\mathcal{B}$ and $\mathcal{B}'$ are Boolean algebras, and if a one-to-one correspondence from $\mathcal{B}$ onto $\mathcal{B}'$ preserves the ordering relation, then this correspondence also preserves least element, greatest element, sup, inf, and complement.

#8. Prove that two finite Boolean algebras are isomorphic if and only if they have the same number of atoms.

9. Go through the steps in the proofs of Lemmas 3.3 through 3.7 for the special case of the Boolean algebra $\mathcal{B}_2$ of all Boolean functions of two independent variables.

5-4 $\mathcal{B}_n$ as a Boolean Algebra; Applications to Logic

We know already that $\mathcal{B}_n$ is a Boolean algebra (Example 2.3). Problem 1 is concerned with the tabular representation of the sup, inf, or complement of Boolean functions whose tabular representations are given.

We have seen in earlier chapters that the correspondence between truth value functions and their associated Boolean functions can be used to transcribe statements about the algebraic structure of $\mathcal{B}_n$ in terms of logical connectives between truth value functions. Any algebraic theorem which holds in all Boolean algebras must certainly hold in the particular Boolean algebra $\mathcal{B}_n$ and should, therefore, give rise to a theorem of logic (tautology). The process is illustrated in the following examples.

Example 4.1 For any elements A, B of a Boolean algebra $\mathcal{B}$,

$$A \cap B \leq A \leq A \cup B.$$

Thus, for any elements ϕ, ψ of the Boolean algebra $\mathcal{B}_n$

$$\phi \cap \psi \leq \phi \leq \phi \cup \psi.$$

These relations, when transcribed in terms of the truth value functions which correspond to ϕ and ψ, give rise to the following tautologies. For any truth value functions f and g,

$$f \wedge g \Rightarrow f \quad \text{and} \quad f \Rightarrow f \vee g.$$

Example 4.2 For any element B of a Boolean algebra $\mathcal{B}$,

$$B \cup \bar{B} = G.$$

Writing this equation successively in terms of elements of $\mathcal{B}_n$, and then in terms of truth value functions, we find: For any truth value function f,

$$f \vee (\sim f) \Leftrightarrow 1.$$

PROBLEMS

1. Suppose that the tabular representations are given for two Boolean functions ϕ and ψ of the same independent variables. Explain how to obtain the tabular representations of $\phi \cup \psi$, $\phi \cap \psi$, and $\bar{\phi}$.

2. Each of the following theorems is valid for elements of an arbitrary Boolean algebra $\mathcal{B}$. Obtain a theorem of logic from each of them.

   (a) $\bar{\bar{A}} = A$.

   (b) $\overline{A \cup B} = \bar{A} \cap \bar{B}$.

   (c) $\overline{A \cap B} = \bar{A} \cup \bar{B}$.

   (d) $A \cap (B \cup C) = (A \cap B) \cup (A \cap C)$.

   (e) $A \cup (B \cap C) = (A \cup B) \cap (A \cup C)$.

#3. For any element A of a Boolean algebra $\mathcal{B}$, $A \leq G$. Obtain a theorem of logic from this information. Contrast your work with that in Examples 4.1 and 4.2.

4. Obtain theorems of logic from the results of Problem 6, Section 5-2.

5. Obtain a theorem of logic from each of the following.

   (a) Theorem 2.2, Section 5-2.

   (b) Theorem 2.5, parts (d) and (e), Section 5-2.

   (c) Lemma 3.3, Section 5-3.

   (d) Lemma 3.4, Section 5-3.

   (e) Lemma 3.5, Section 5-3.

6

BOOLEAN RINGS

6-1 Introduction

A ring is an algebraic system composed of a set, together with two binary operations (called addition and multiplication) on the set, in which certain postulates are satisfied. There are several different (but equivalent) sets of postulates which can be used to characterize a ring. In Section 6-2, one of these sets of postulates is used to define a ring and some of the basic properties of a ring are derived. The set of postulates which we use to define a ring is a little different from the set which is most frequently used. Thus, even for a student who has some previous acquaintance with rings, the material in Section 6-2 should be of interest since the method of approach will probably be different from that with which the student is familiar. A Boolean ring is defined, and the problems of Section 6-2 provide some practice in the techniques and methods of proof in the elementary theory of rings. Section 6-3 discusses Boolean rings with a unit element. These algebraic systems appear, at first glance, to have very little in common with Boolean algebras, but the main result of Section 6-3 is that these two types of algebraic systems are equivalent in the sense that any Boolean algebra can be thought of as a Boolean ring with a unit element and every Boolean ring with a unit element can be thought of as a Boolean algebra. In Section 6-4, it is shown that every finite Boolean ring can be represented in a particular way as a Boolean ring whose elements are n-tuples of Boolean constants.

6-2 Definitions and Basic Properties

A ring is an algebraic system which is intermediate between two others (group and field) in the sense that every field is a ring and every ring is a group. The study of rings is usually preceded either by a study of groups or by a study of fields. Thus, in the usual presentation, a ring appears either as a generalization of a field or as a specialization of a group. Each of these algebraic systems is of major importance both in pure and in applied mathematics, but neither groups nor fields are required for our purposes in this chapter. We therefore confine our attention to rings and define this algebraic system directly.

DEFINITION 2.1 A *ring* is a non-empty set $\mathcal{R}$ in which two binary operations "$+$" and "$\cdot$" are defined satisfying the following conditions: If A, B, C are any elements of $\mathcal{R}$, then

(a) Closure: $A + B$ and $A \cdot B$ are unique elements of $\mathcal{R}$.

(b) Commutativity of "$+$": $A + B = B + A$.

(c) Associativity of "$+$": $A + (B + C) = (A + B) + C$.

(d) Solvability of Equations: The equation $A + X = B$ has at least one solution in $\mathcal{R}$.

(e) Associativity of "$\cdot$": $A \cdot (B \cdot C) = (A \cdot B) \cdot C$.

(f) Distributivity:

$$A \cdot (B + C) = A \cdot B + A \cdot C$$
$$(A + B) \cdot C = A \cdot C + B \cdot C.$$

Several remarks are in order concerning Definition 2.1. The operations "$+$" and "$\cdot$" will be called addition and multiplication respectively. As is customary with ordinary arithmetical multiplication, the "$\cdot$" will sometimes be omitted. Thus "$A \cdot B$" may be written as "AB". According to Definition 2.1, addition and multiplication enjoy many of the properties of the ordinary arithmetical operations. However, it is important to notice that some properties are not mentioned. For example, multiplication is not required to be commutative and there is no explicit mention of special elements with properties analogous to those of the numbers 0 and 1. We shall prove, as a theorem, that the analog of zero must exist in every ring, but there may be no analog of the number 1, as is shown by Example 2.2 below. Condition (a) of Definition 2.1 is really included in the statement that addition and multiplication are binary operations defined on $\mathcal{R}$, but it is stated explicitly because of its importance in some of our proofs. Note also that each of conditions (a) through (f) is to hold for arbitrary elements A, B, C of $\mathcal{R}$. For example, condition (d) may be stated: For any elements A and B of $\mathcal{R}$, there is at least one element X of $\mathcal{R}$ such that $A + X = B$.

Example 2.2 The set of all even integers (positive, negative, and zero) is a ring with the usual arithmetical operations of addition and multiplication.

Example 2.3 The set of all 2×2 matrices is a ring with matrix addition and multiplication.

The student familiar with the rudiments of matrix theory will recognize that, for every positive integer n, the set of all $n \times n$ matrices is a ring with matrix addition and multiplication. The special case of the 2×2 matrices will suffice for the purposes of this example. In this special case, the definitions are as follows. A 2×2 *matrix* A is a square array of 4 numbers.

$$A = \begin{pmatrix} a & b \\ c & d \end{pmatrix}$$

The numbers a, b, c, d are called the *elements* of the matrix A. Two matrices are equal if their elements are respectively equal. Matrix addition and multiplication are defined by

$$\begin{pmatrix} a & b \\ c & d \end{pmatrix} + \begin{pmatrix} e & f \\ g & h \end{pmatrix} = \begin{pmatrix} a + e & b + f \\ c + g & d + h \end{pmatrix}$$

and

$$\begin{pmatrix} a & b \\ c & d \end{pmatrix} \cdot \begin{pmatrix} e & f \\ g & h \end{pmatrix} = \begin{pmatrix} ae + bg & af + bh \\ ce + dg & cf + dh \end{pmatrix}.$$

Example 2.4 The set $\{0, 1\}$ of Boolean constants is a ring with the operations of addition and multiplication defined in the following tables.

+	0	1
0	0	1
1	1	0

·	0	1
0	0	0
1	0	1

Example 2.5 The set of all n-tuples of Boolean constants is a ring with the operations of addition and multiplication defined by

$$(\alpha_1, \alpha_2, \ldots, \alpha_n) + (\beta_1, \beta_2, \ldots, \beta_n) = (\alpha_1 + \beta_1, \alpha_2 + \beta_2, \ldots, \alpha_n + \beta_n)$$

and

$$(\alpha_1, \alpha_2, \ldots, \alpha_n) \cdot (\beta_1, \beta_2, \ldots, \beta_n) = (\alpha_1 \cdot \beta_1, \alpha_2 \cdot \beta_2, \ldots, \alpha_n \cdot \beta_n).$$

In these equations the operations "+" and "·" appearing in the right members are to be interpreted as those of Example 2.4. For example,

$$(1, 0, 0) + (1, 1, 0) = (0, 1, 0)$$

and

$$(1, 0, 0) \cdot (1, 1, 0) = (1, 0, 0).$$

The next two theorems show that, in any ring there is an element with properties analogous to those of the number zero.

THEOREM 2.6 In any ring $\mathcal{R}$ there is a unique element 0 such that, for all elements A of $\mathcal{R}$, $A + 0 = A$.

Proof. It is easy to see that there cannot be two different elements, both of which satisfy the conditions of the theorem. If 0_1 and 0_2 each satisfy these conditions, then

$$0_1 = 0_1 + 0_2 = 0_2 + 0_1 = 0_2.$$

Thus the element 0 is unique, provided it exists.

To prove existence, choose any element A_0 in $\mathcal{R}$. By condition (d) of Definition 2.1, the equation

$$A_0 + X = A_0$$

has at least one solution. Let 0 be any solution of this equation, and let B be any element in $\mathcal{R}$. We must show that $B + 0 = B$. Again applying condition (d) of Definition 2.1, we find an element C of $\mathcal{R}$ such that

$$A_0 + C = B.$$

Then

$$B + 0 = (A_0 + C) + 0 = C + (A_0 + 0) = C + A_0 = B. \blacksquare$$

Since the element 0 of $\mathcal{R}$ whose existence was proved in Theorem 2.6 has properties analogous to those of the number zero, no confusion should arise from denoting them both by the symbol "0". However, the reader is reminded that the symbol "0" has been endowed with several different meanings, and a little care will be required to keep all these meanings properly sorted out.

The proof of Theorem 2.6 shows that, for any element A_0 of $\mathcal{R}$, the equation $A_0 + X = A_0$ has a unique solution. This remark will be used in the proof of the following theorem.

THEOREM 2.7 For any element A of a ring $\mathcal{R}$,

$$A \cdot 0 = 0 = 0 \cdot A.$$

Proof. Since

$$A + 0 = A,$$

we may multiply by A, first on the left, and then on the right, and use the distributive law to obtain

$$A \cdot A + A \cdot 0 = A \cdot A$$

and

$$A \cdot A + 0 \cdot A = A \cdot A.$$

Thus each of $A \cdot 0$ and $0 \cdot A$ is a solution of the equation

$$A \cdot A + X = A \cdot A.$$

By the remark immediately preceding this theorem,

$$A \cdot 0 = 0 = 0 \cdot A. \blacksquare$$

(Caution: It can happen that $AB = 0$ even when $A \neq 0$ and $B \neq 0$; see Problem 2b.)

We close this section with three definitions. The exercises which follow will provide some acquaintance with the fundamental properties of these concepts.

DEFINITION 2.8 An *additive inverse* of the element A of a ring $\mathcal{R}$ is any solution of the equation $A + X = 0$.

It follows from Problem 6 that each element in a ring has a unique additive inverse. We shall denote the additive inverse of A by "$-A$".

DEFINITION 2.9 A *unit element* of a ring $\mathcal{R}$ is an element 1 of $\mathcal{R}$ (if it exists) such that, for each element A of $\mathcal{R}$,

$$A \cdot 1 = A = 1 \cdot A.$$

A ring may have a unit element, or it may not. Problem 1 asks for examples of rings with and without a unit element. Since the definition of a unit element requires it to have properties analogous to those of the number one, we shall use the symbol "1" for a unit element in a ring.

DEFINITION 2.10 A *Boolean ring* is a ring $\mathcal{R}$ such that, for each element A of $\mathcal{R}$,

$$A \cdot A = A.$$

PROBLEMS

1. Find two examples of a ring with a unit element and two examples of a ring with no unit element.

2. (a) Can there be two different unit elements in a ring?

   (b) Give an example of two elements A, B in a ring $\mathcal{R}$ such that $A \neq 0$, $B \neq 0$, $AB = 0$. Can $\mathcal{R}$ be a Boolean ring?

3. Find two examples of a Boolean ring.

4. Find an example of a Boolean ring with no unit element.

5. Prove that, for any element A of a Boolean ring $\mathcal{R}$,

$$A^n = A.$$

   (The exponent n is a positive integer and indicates successive multiplication.)

6. For any elements A, B of a ring $\mathcal{R}$, prove that the equation

$$A + X = B$$

   has a unique solution in $\mathcal{R}$. The solution of this equation is denoted by $B - A$.

7. Prove that, for any elements A, B of a ring $\mathcal{R}$,

$$A - B = A + (-B).$$

   (The "$-$" on the left member is the operation of Problem 6; the "$-$" in the right member denotes the additive inverse; this result shows that these two operations in an arbitrary ring are connected in the same way as the analogous operations on real numbers. Thus, no confusion need result from using the same symbol to denote these two different operations.)

#8. Prove that, for any elements A, B of a Boolean ring $\mathcal{R}$,

$$AB + BA = 0.$$

[*Hint:* Consider the expression $(A + B)(A + B)$.]

#9. For any element A of a Boolean ring $\mathcal{R}$, show that

$$A = -A.$$

That is, each element in a Boolean ring is its own additive inverse.

#10. Show that, in a Boolean ring, multiplication is commutative.

11. Simplify the expression

$$A - (A^2 + B - A^3)(C^3 + AC^2 - A^2C) + ABC(A + B + C)$$

where A, B, C are elements of a Boolean ring.

12. (a) Show that there is a Boolean ring with exactly one element, and that any two Boolean rings, with one element each, are isomorphic.

(b) Show that there is a Boolean ring with exactly two elements, and that any two Boolean rings with two elements each, are isomorphic.

(c) Show that there is no Boolean ring with exactly three elements, but show that there is a ring with three elements.

6-3 Boolean Rings with a Unit Element are Equivalent to Boolean Algebras

The two theorems of this section prove the equivalence of the algebraic systems mentioned in the title of the section. We shall prove that, in any Boolean algebra $\mathcal{B}$, it is possible to define two binary operations "+" and "·" so that $\mathcal{B}$ becomes a Boolean ring with a unit element. Conversely, in any Boolean ring $\mathcal{R}$ with a unit element, it is possible to define an ordering relation so that $\mathcal{R}$ becomes a Boolean algebra. Some further properties of this equivalence are mentioned in the problems.

THEOREM 3.1 Let $\mathcal{B}$ be a Boolean algebra. Then $\mathcal{B}$ becomes a Boolean ring with a unit element if the binary operations "+" and "·" are defined in $\mathcal{B}$ by

$$A + B = (A \cup B) \cap (\bar{A} \cup \bar{B})$$

and

$$A \cdot B = A \cap B.$$

Proof. We must show that the operations "+" and "·" satisfy the conditions of Definitions 2.1 and 2.10, and that there is a unit element. We shall prove here only conditions (c) and (d) of Definition 2.1. The remainder of the proof is left as an exercise (Problem 1).

Associativity of "+": From the definition of "+" in Theorem 3.1, we find

$$A + (B + C)$$
$$= \{A \cup [(B \cup C) \cap (\bar{B} \cup \bar{C})]\} \cap \{\bar{A} \cup \overline{[(B \cup C) \cap (\bar{B} \cup \bar{C})]}\}.$$

Using the distributivity of "$\cup$" over "$\cap$" and De Morgan's Laws, the right member may be written as

$$(A \cup B \cup C) \cap (A \cup \bar{B} \cup \bar{C}) \cap \{\bar{A} \cup [(\bar{B} \cap \bar{C}) \cup (B \cap C)]\}.$$

Now we apply the associative law inside the curly bracket, and again distribute "$\cup$" over "$\cap$" to obtain

$$(A \cup B \cup C) \cap (A \cup \bar{B} \cup \bar{C}) \cap \{[(\bar{A} \cup \bar{B}) \cap (\bar{A} \cup \bar{C})] \cup (B \cap C)\}.$$

We apply the distributive law to the material in the curly bracket twice more in succession. The first application gives

$$(A \cup B \cup C) \cap (A \cup \bar{B} \cup \bar{C})$$
$$\cap \{[(\bar{A} \cup \bar{B}) \cup (B \cap C)] \cap [(\bar{A} \cup \bar{C}) \cup (B \cap C)]\}$$

and the second application results in

$$(A \cup B \cup C) \cap (A \cup \bar{B} \cup \bar{C}) \cap (\bar{A} \cup \bar{B} \cup B)$$
$$\cap (\bar{A} \cup \bar{B} \cup C) \cap (\bar{A} \cup \bar{C} \cup B) \cap (\bar{A} \cup \bar{C} \cup C).$$

Since

$$(\bar{A} \cup \bar{B} \cup B) = G \quad \text{and} \quad (\bar{A} \cup \bar{C} \cup C) = G,$$

these two expressions may be dropped from this repeated inf; we have shown

$$A + (B + C)$$
$$= (A \cup B \cup C) \cap (A \cup \bar{B} \cup \bar{C}) \cap (\bar{A} \cup \bar{B} \cup C) \cap (\bar{A} \cup B \cup \bar{C}).$$

The expression $(A + B) + C$ is treated similarly. The successive results are as follows.

$$(A + B) + C$$
$$= \{[(A \cup B) \cap (\bar{A} \cup \bar{B})] \cup C\} \cap \{\overline{[(A \cup B) \cap (\bar{A} \cup \bar{B})]} \cup \bar{C}\}$$
$$= (A \cup B \cup C) \cap (\bar{A} \cup \bar{B} \cup C) \cap \{[(\bar{A} \cap \bar{B}) \cup (A \cap B)] \cup \bar{C}\}$$
$$= (A \cup B \cup C) \cap (\bar{A} \cup \bar{B} \cup C) \cap \{(\bar{A} \cap \bar{B}) \cup [(A \cup \bar{C}) \cap (B \cup \bar{C})]\}$$

$$= (A \cup B \cup C) \cap (\bar{A} \cup \bar{B} \cup C) \cap \{[(\bar{A} \cap \bar{B}) \cup (A \cup \bar{C})] \cap [(\bar{A} \cap \bar{B}) \cup (B \cup \bar{C})]\}$$

$$= (A \cup B \cup C) \cap (\bar{A} \cup \bar{B} \cup C) \cap (\bar{A} \cup A \cup \bar{C}) \cap (\bar{B} \cup A \cup \bar{C}) \cap (\bar{A} \cup B \cup \bar{C}) \cap (\bar{B} \cup B \cup \bar{C})$$

$$= (A \cup B \cup C) \cap (\bar{A} \cup \bar{B} \cup C) \cap (A \cup \bar{B} \cup \bar{C}) \cap (\bar{A} \cup B \cup \bar{C}).$$

Thus

$$A + (B + C) = (A + B) + C.$$

Solvability of equations: If A and B are any elements of $\mathcal{B}$, we shall show that the element

$$X = (A \cup B) \cap (\bar{A} \cup \bar{B})$$

is a solution of the equation $A + X = B$. By direct computation, we find

$$A + [(A \cup B) \cap (\bar{A} \cup \bar{B})]$$

$$= \{A \cup [(A \cup B) \cap (\bar{A} \cup \bar{B})]\} \cap \{\bar{A} \cup \overline{[(A \cup B) \cap (\bar{A} \cup \bar{B})]}\}$$

$$= \{(A \cup A \cup B) \cap (A \cup \bar{A} \cup \bar{B})\} \cap \{\bar{A} \cup (\bar{A} \cap \bar{B}) \cup (A \cap B)\}$$

$$= (A \cup B) \cap G \cap \{\bar{A} \cup (A \cap B)\}$$

$$= (A \cup B) \cap (\bar{A} \cup A) \cap (\bar{A} \cup B)$$

$$= (A \cup B) \cap (\bar{A} \cup B) = (A \cap \bar{A}) \cup B = B. \blacksquare$$

THEOREM 3.2 Let $\mathcal{R}$ be a Boolean ring with a unit element. Then $\mathcal{R}$ becomes a Boolean algebra if the ordering relation "$\leq$" is defined in $\mathcal{R}$ by

$$A \leq B \Leftrightarrow A \cdot B = A.$$

Proof. It is evident that "$\leq$" is a relation on $\mathcal{R}$ since, for any elements A, B of $\mathcal{R}$, the expression

$$A \cdot B = A$$

is a sentence. To prove that "$\leq$" is an ordering relation, note that, if A is any element of $\mathcal{R}$, then $A \cdot A = A$. Thus

$$A \leq A \Leftrightarrow 1,$$

and "$\leq$" is reflexive. If $A \leq B$ and $B \leq A$, then

$$A = A \cdot B = B \cdot A = B.$$

Thus "$\leq$" is antisymmetric. If $A \leq B$ and $B \leq C$, then

$$A \cdot C = (A \cdot B) \cdot C = A \cdot (B \cdot C) = A \cdot B = A.$$

Thus "$\leq$" is transitive. Since "$\leq$" is reflexive, antisymmetric, and transitive, it is an ordering relation.

To show that the ordered set $\mathcal{R}$ is a lattice, it suffices to prove

$$\sup \{A, B\} = A + B + AB$$

and

$$\inf \{A, B\} = AB.$$

We find

$$A(A + B + AB) = AA + AB + AAB = A + AB + AB.$$

But $AB + AB = 0$ since each element of a Boolean ring is its own additive inverse. Thus

$$A(A + B + AB) = A.$$

Similarly,

$$B(A + B + AB) = B.$$

Hence

$$A \leq A + B + AB \quad \text{and} \quad B \leq A + B + AB;$$

FIGURE 3.1 Correspondence between Operations, Relations, and Special Elements of a Boolean Algebra and a Boolean Ring with a Unit Element.

Boolean Algebra	Boolean Ring with a Unit Element
$A \cup B$	$A + B + AB$
$A \cap B$	AB
$\bar{A}$	$1 + A$
$A \leq B$	$AB = A$
$(A \cup B) \cap (\bar{A} \cup \bar{B})$	$A + B$
A	$-A$
G	1
L	0

that is, $A + B + AB$ is an upper bound for $\{A, B\}$. If C is any upper bound for $\{A, B\}$, then $AC = A$ and $BC = B$, thus

$$(A + B + AB)C = AC + BC + ABC = A + B + AB$$

so that

$$A + B + AB \leq C,$$

proving that $A + B + AB = \sup \{A, B\}$. The proof that $\inf \{A, B\} = AB$ is similar, and is left as an exercise. (Problem 2).

It is easy to see that 1 and 0 are, respectively, the greatest and least elements in $\mathcal{R}$ since, for any element A in $\mathcal{R}$,

$$A \cdot 1 = A \quad \text{and} \quad 0 \cdot A = 0.$$

Moreover, $1 + A$ is a complement of A since

$$A \cap (1 + A) = A(1 + A) = A + A = 0$$

and

$$\begin{aligned} A \cup (1 + A) &= A + (1 + A) + A(1 + A) \\ &= A + 1 + A + A + A \\ &= 1 + (A + A) + (A + A) \\ &= 1 + 0 + 0 = 1. \end{aligned}$$

The proof that "$\cap$" distributes over "$\cup$" is left as an exercise (Problem 2). Since $\mathcal{R}$ is a complemented, distributive lattice, it is a Boolean algebra. ∎

For reference purposes, we give a table (Fig. 3.1) showing the correspondence set up by Theorems 3.1 and 3.2 between the operations, relations, and the special elements of a Boolean algebra and a Boolean ring with a unit element.

PROBLEMS

1. (a) Complete the proof of Theorem 3.1 by showing that the remaining conditions of Definitions 2.1 and 2.10 are satisfied, and that there is a unit element. Which element of $\mathcal{B}$ becomes the element 0 of the Boolean ring?

   (b) Supply a reason for the validity of each of the steps in rewriting the expression $(A + B) + C$ in the proof of Theorem 3.1.

(c) Try to discover some system in the procedure for rewriting $A + (B + C)$ and $(A + B) + C$ in the proof of Theorem 3.1. Why were the steps performed in the particular order shown in the text?

(d) Provide a reason for the validity of each of the steps in the proof of the solvability of equations in Theorem 3.1.

#2. (a) Prove that $\inf\{A, B\} = AB$ as stated in the proof of Theorem 3.2.

(b) Complete the proof of Theorem 3.2 by showing that the lattice $\mathcal{R}$ is distributive.

3. Check that the corresponding operations and elements are given correctly in Figure 3.1.

4. (a) Write the expression $A \cup (B \cap C) \cup D$ in terms of addition and multiplication.

(b) Write the expression $A + [B - C(A + B)]$ in terms of "$\cup$", "$\cap$", and "$-$".

#5. Prove that every finite Boolean ring has a unit element. (*Hint:* Use the ideas in the first part of the proof of Theorem 3.2.)

6. The correspondences set up in Theorems 3.1 and 3.2 are reciprocal to each other in a certain sense as indicated in the two parts of this problem.

(a) Let $\mathcal{B}$ be a Boolean algebra and let $\mathcal{R}$ be the Boolean ring with a unit element obtained by defining "$+$" and "$\cdot$" in $\mathcal{B}$ as in Theorem 3.1. Show that the ordering relation defined in $\mathcal{R}$ as in Theorem 3.2 is the original ordering relation in $\mathcal{B}$.

(b) Let $\mathcal{R}$ be a Boolean ring with a unit element and let $\mathcal{B}$ be the Boolean algebra obtained by defining an ordering relation "$\leq$" in $\mathcal{R}$ as in Theorem 3.2. Show that the binary operations "$+$" and "$\cdot$" defined in $\mathcal{B}$ as in Theorem 3.1 are the original addition and multiplication in $\mathcal{R}$, respectively.

#7. Apply the definitions of the binary operations "$+$" and "$\cdot$"; as given in Theorem 3.1, to the special case of the Boolean algebra $\mathcal{B}_n$ of all Boolean functions of n independent variables. Show that, for any elements ϕ and ψ of $\mathcal{B}_n$, and for any Boolean constants $\alpha_1, \alpha_2, \ldots, \alpha_n$,

$$(\phi + \psi)(\alpha_1, \alpha_2, \ldots, \alpha_n) = \begin{cases} 1, & \text{iff } \phi(\alpha_1, \alpha_2, \ldots, \alpha_n) \\ & \quad \neq \psi(\alpha_1, \alpha_2, \ldots, \alpha_n) \\ 0, & \text{iff } \phi(\alpha_1, \alpha_2, \ldots, \alpha_n) \\ & \quad = \psi(\alpha_1, \alpha_2, \ldots, \alpha_n) \end{cases}$$

$$(\phi\cdot\psi)(\alpha_1, \alpha_2, \ldots, \alpha_n) = \begin{cases} 1, & \text{iff } \phi(\alpha_1, \alpha_2, \ldots, \alpha_n) = 1 \\ & \text{and } \psi(\alpha_1, \alpha_2, \ldots, \alpha_n) = 1 \\ 0, & \text{iff } \phi(\alpha_1, \alpha_2, \ldots, \alpha_n) = 0 \\ & \text{or } \psi(\alpha_1, \alpha_2, \ldots, \alpha_n) = 0. \end{cases}$$

#8. (a) Prove that if we define the operations "+" and "$\cdot$" in $\mathcal{B}_n$ as in Problem 7, and define the operations "+" and "$\cdot$" between Boolean constants as in Example 2.4, then, for any elements ϕ and ψ of $\mathcal{B}_n$, and for any Boolean constants $\alpha_1, \alpha_2, \ldots, \alpha_n$,

$$(\phi + \psi)(\alpha_1, \alpha_2, \ldots, \alpha_n) = \phi(\alpha_1, \alpha_2, \ldots, \alpha_n) + \psi(\alpha_1, \alpha_2, \ldots, \alpha_n)$$

and

$$(\phi\cdot\psi)(\alpha_1, \alpha_2, \ldots, \alpha_n) = \phi(\alpha_1, \alpha_2, \ldots, \alpha_n)\cdot\psi(\alpha_1, \alpha_2, \ldots, \alpha_n).$$

(b) If two Boolean functions ϕ and ψ are given by their tabular representations, explain how to obtain the tabular representations for the Boolean functions $\phi + \psi$ and $\phi\cdot\psi$.

6-4 Representation of a Finite Boolean Ring in Terms of n-tuples of Boolean Constants

In Problems 7 and 8, Section 6-3, we have gained some acquaintance with the Boolean ring formed by the set $\mathcal{B}_n$ of all Boolean functions of n independent variables. A Boolean function of n independent variables may be thought of as a 2^n-tuple of Boolean constants — namely, the 2^n entries in the tabular representation of the function. As we have seen in Example 2.5, the set of all n-tuples of Boolean constants also forms a Boolean ring. The main result of this section is that these are essentially the only finite Boolean rings; every finite Boolean ring is isomorphic to the ring of all n-tuples of Boolean constants for some n.

THEOREM 4.1 If $\mathcal{R}$ is any finite Boolean ring, then (for some non-negative integer n) $\mathcal{R}$ is isomorphic to the Boolean ring of all n-tuples of Boolean constants.

Proof. The proof consists of Lemmas 4.2 through 4.7 below. We shall use the following notation in these lemmas.

$\mathcal{R}$: a (given) finite Boolean ring (with a unit element, by Problem 5 Section 6-3).

$\mathcal{B}$: the Boolean algebra obtained from $\mathcal{R}$ by the definition

$$A \leq B \Leftrightarrow AB = A.$$

The operations "$\cup$", "$\cap$", and "$^{-}$" in the Boolean algebra $\mathcal{B}$ will be used in the proof as well as the operations "$+$" and "$\cdot$" in the ring $\mathcal{R}$.
$\mathcal{A} = \{A_1, A_2, \ldots, A_n\}$: the set of all atoms in $\mathcal{B}$; that is, all elements A of $\mathcal{B}$ such that $X \leq A$ has exactly two solutions. Note that $A_1, A_2, \ldots, A_n$ are atoms; we shall use A or B for an arbitrary element of $\mathcal{B}$.

At some places in the proof we shall assume that $\mathcal{R}$ has at least two elements; the case where $\mathcal{R}$ contains exactly one element is left as an exercise (Problem 1).

LEMMA 4.2 For any atoms A_i, A_j of $\mathcal{B}$,

$$A_i \cdot A_j = \begin{cases} A_i, & \text{if } A_i = A_j. \\ 0, & \text{if } A_i \neq A_j. \end{cases}$$

Proof. From Problem 2, Section 6-3,.

$$A_i \cdot A_j = A_i \cap A_j.$$

Evidently, if $A_i = A_j$, then $A_i \cap A_j = A_i$. If $A_i \neq A_j$ then, since each of A_i and A_j is an atom, the only lower bound of $\{A_i, A_j\}$ is 0. Thus

$$A_i \cap A_j = 0. \blacksquare$$

LEMMA 4.3 For any atom A_i,

$$A_i \leq A_1 + A_2 + \ldots + A_n.$$

Proof. From Lemma 4.2,

$$A_i \cdot (A_1 + A_2 + \ldots + A_n) = 0 + 0 + \ldots + A_i + \ldots + 0 = A_i.$$

Thus, from the definition of the ordering relation "$\leq$",

$$A_i \leq A_1 + A_2 + \ldots + A_n. \blacksquare$$

LEMMA 4.4 For any element $B \neq 0$ of $\mathcal{R}$, there is an atom A_i such that $A_i \leq B$.

Proof. This is just a restatement of Lemma 3.4, Section 5-3. ∎

LEMMA 4.5 The sum of all the atoms in $\mathcal{B}$ is the unit element in $\mathcal{R}$; that is

$$A_1 + A_2 + \ldots + A_n = 1.$$

Proof. Set

$$B = \overline{A_1 + A_2 + \ldots + A_n}.$$

We shall show that, for each atom A_i,

$$A_i \leq B \Leftrightarrow 0.$$

It will follow from Lemma 4.4 that $B = 0$, and that

$$A_1 + A_2 + \ldots + A_n = \bar{B} = 1.$$

The proof is by contradiction. If it were true that $A_i \leq B$, then

$$\overline{A_i} \geq \bar{B} = A_1 + A_2 + \ldots + A_n.$$

It would follow that

$$A_i \cap \overline{A_i} \geq A_i \cap (A_1 + A_2 + \ldots + A_n).$$

But $A_i \cap \overline{A_i} = L$, whereas, by Lemma 4.3,

$$A_i \cap (A_1 + A_2 + \ldots + A_n) = A_i,$$

which would lead to the contradiction

$$L \geq A_i. \blacksquare$$

LEMMA 4.6 For any element A of $\mathcal{R}$, there is a unique n-tuple of Boolean constants $(\alpha_1, \alpha_2, \ldots, \alpha_n)$ such that

$$A = \alpha_1 A_1 + \alpha_2 A_2 + \ldots + \alpha_n A_n.$$

Proof. For each $i (1 \leq i \leq n)$, $A \cap A_i \leq A_i$. Thus $A \cap A_i$ is either 0 or A_i. Set

$$\alpha_i = \begin{cases} 0, & \text{if } A \cdot A_i = 0. \\ 1, & \text{if } A \cdot A_i = A_i. \end{cases} \qquad (i = 1, 2, \ldots, n)$$

Then

$$A = A \cdot 1 = A(A_1 + A_2 + \ldots + A_n) = \alpha_1 A_1 + \alpha_2 A_2 + \ldots + \alpha_n A_n.$$

If

$$\alpha_1 A_1 + \alpha_2 A_2 + \ldots + \alpha_n A_n = A = \beta_1 A_1 + \beta_2 A_2 + \ldots + \beta_n A_n,$$

then, for each i $(1 \leq i \leq n)$,

$$\begin{aligned} \alpha_i A_i &= A_i(\alpha_1 A_1 + \alpha_2 A_2 + \ldots + \alpha_n A_n) \\ &= A_i(\beta_1 A_1 + \beta_2 A_2 + \ldots + \beta_n A_n) = \beta_i A_i. \end{aligned}$$

Hence $\alpha_i = \beta_i$, and the n-tuple of Boolean constants $(\alpha_1, \alpha_2, \ldots, \alpha_n)$ is uniquely determined by the element A and the conditions of the lemma. ▮

LEMMA 4.7 The correspondence of Lemma 4.6 between an arbitrary element A of $\mathcal{R}$ and an n-tuple of Boolean constants $(\alpha_1, \alpha_2, \ldots, \alpha_n)$ is an isomorphism between $\mathcal{R}$ and $\mathcal{R}_n$, where $\mathcal{R}_n$ is the Boolean ring of all n-tuples of Boolean constants.

Proof. By Lemma 4.6, the correspondence is one-to-one between $\mathcal{R}$ and some subset of $\mathcal{R}_n$. Evidently it is onto $\mathcal{R}_n$ since, if $(\alpha_1, \alpha_2, \ldots, \alpha_n)$ is any n-tuple of Boolean constants, then

$$A = \alpha_1 A_1 + \alpha_2 A_2 + \ldots + \alpha_n A_n$$

is an element of $\mathcal{R}$ which corresponds to this particular n-tuple.

To complete the proof, we must show that the correspondence preserves the ring operations "+" and "·" in $\mathcal{R}$ and $\mathcal{R}_n$. If A and B are any elements of $\mathcal{R}$, and if

$$A = \alpha_1 A_1 + \alpha_2 A_2 + \ldots + \alpha_n A_n$$

and

$$B = \beta_1 A_1 + \beta_2 A_2 + \ldots + \beta_n A_n,$$

then

$$A + B = (\alpha_1 + \beta_1)A_1 + (\alpha_2 + \beta_2)A_2 + \ldots + (\alpha_n + \beta_n)A_n$$

where, for each i $(1 \leq i \leq n)$,

$$\alpha_i + \beta_i = \begin{cases} 1, & \text{if } \alpha_i \neq \beta_i \\ 0, & \text{if } \alpha_i = \beta_i \end{cases}$$

in accordance with the table in Example 2.4, Section 6-2. Also,

$$\begin{aligned} A \cdot B &= (\alpha_1 A_1 + \alpha_2 A_2 + \ldots + \alpha_n A_n)(\beta_1 A_1 + \beta_2 A_2 + \ldots + \beta_n A_n) \\ &= \alpha_1 \beta_1 A_1 + \alpha_2 \beta_2 A_2 + \ldots + \alpha_n \beta_n A_n \end{aligned}$$

where, for each i $(1 \leq i < n)$,

$$\alpha_i \beta_i = \begin{cases} 1, & \text{if } \alpha_i = 1 = \beta_i \\ 0, & \text{otherwise} \end{cases}$$

in accordance with the table in Example 2.4, Section 6-2. ▮

This completes the proof of Theorem 4.1. ▮

Note that we have proved that the integer n mentioned in Theorem 4.1 is the number of atoms in the Boolean algebra $\mathcal{B}$.

PROBLEMS

1. (a) Prove Theorem 4.1 in the special case where $\mathcal{R}$ has exactly one element. Is the proof in the text valid for this case? If not, where does it break down?

   (b) In this special case, is it again true that the integer n mentioned in the theorem is the number of atoms in the Boolean algebra $\mathcal{B}$ constructed in the proof?

2. (a) Prove that, for every finite Boolean ring $\mathcal{R}$, the number of elements in $\mathcal{R}$ is a power of 2.

FIGURE 4.1

+	a	b	c	d	e	f	g	h
a	a	b	c	d	e	f	g	h
b	b	a	d	c	f	e	h	g
c	c	d	a	b	h	g	f	e
d	d	c	b	a	g	h	e	f
e	e	f	h	g	a	b	d	c
f	f	e	g	h	b	a	c	d
g	g	h	f	e	d	c	a	b
h	h	g	e	f	c	d	b	a

$\cdot$	a	b	c	d	e	f	g	h
a	a	a	a	a	a	a	a	a
b	a	b	b	a	b	a	b	a
c	a	b	c	d	c	d	b	a
d	a	a	d	d	d	d	a	a
e	a	b	c	d	e	f	g	h
f	a	a	d	d	f	f	h	h
g	a	b	b	a	g	h	g	h
h	a	a	a	a	h	h	h	h

(b) Prove also that, for every non-negative integer n, there is a Boolean ring with 2^n elements.

3. **(a)** Prove that the set of all n-tuples of Boolean constants, with the relation "$\leq$" defined by

$$(\alpha_1, \alpha_2, \ldots, \alpha_n) \leq (\beta_1, \beta_2, \ldots, \beta_n) \Leftrightarrow \text{For each } i \ (1 \leq i \leq n),$$
$$\alpha_i \text{ is less than or equal to } \beta_i,$$

is a Boolean algebra.

(b) Prove in two ways that each finite Boolean algebra is isomorphic to the Boolean algebra in part (a) (for some non-negative integer n). (*Hint:* A proof can be based on Theorems 3.1 and 4.1; an alternate proof can be based on Theorem 3.1, Section 5-3.)

4. Give an alternative proof of Theorem 4.1 based on Problem 5, Section 6-3, Theorem 3.2, and Problem 3(b).

5. **(a)** Show that, in general, there is more than one isomorphism between a given finite Boolean ring and the Boolean ring of all n-tuples of Boolean constants.

(b) Are there any special cases where this isomorphism is unique?

(c) How many different isomorphisms are there in the general case?

6. Figure 4.1 defines two binary operations "$+$" and "$\cdot$" in the set $\{a, b, c, d, e, f, g, h\}$. The set with these operations is a Boolean ring; find an isomorphism between this ring and the Boolean ring of all n-tuples of Boolean constants for some integer n.

7

NORMAL FORMS, DUALITY

7-1 Introduction

Early in the study of high school algebra, the student learns how to "simplify" an algebraic expression. Sometimes elaborate rules are formulated to tell which of two equivalent expressions is the simpler; sometimes the student is left to decide from his own common sense which of two alternate forms is to be preferred. It is usually in the study of the calculus that the student first appreciates that the simplification of an expression may be both difficult and important, and that two expressions which appear, at first, to be completely unrelated may, in the end, turn out to be identical.

The simplification of expressions in a Boolean algebra is also of importance. As we shall see in the next chapter, such expressions are of interest in the design of electrical networks for various purposes. In this connection, simplification of an expression may lead to a substantial saving in the cost of the electrical hardware used to obtain a particular result. Unfortunately, different viewpoints lead to different criteria for the simplicity of expressions in a Boolean algebra, and, in most cases, there is no simple procedure for reducing a given expression to its simplest form.

Two standard forms for expressions in a Boolean algebra $\mathcal{B}$ are presented in Section 7-2. It will be evident that these forms may not be the simplest possible, and, in general, an element of $\mathcal{B}$ does not uniquely determine either one of these forms. A specialization, to obtain uniqueness, is presented in the problems. This specialized standard form can be used to decide whether or not two given expressions represent the same element.

In Section 7-3 the ring operations "$+$" and "$\cdot$" are used instead of the operations "$\cup$", "$\cap$", and "$^{-}$" of a Boolean algebra. A standard form is presented for expressing an element of a Boolean ring, with a unit element, by applying the ring operations to certain base elements. Again, in general, an element does not uniquely determine its ring standard form.

In Section 7-4, the results of Sections 7-2 and 7-3 are specialized to apply to the Boolean algebra $\mathcal{B}_n$, and it is shown that, in this special case, each element uniquely determines each one of the three standard forms.

Finally, in Section 7-5, the important principle of duality is presented and a few of its applications are indicated.

7-2 Disjunctive and Conjunctive Normal Forms

Let $A_1, A_2, \ldots, A_n$ be n distinct elements (not necessarily atoms) of a Boolean algebra $\mathcal{B}$. We say that an element A of $\mathcal{B}$ is *expressible* in terms of the elements $A_1, A_2, \ldots, A_n$ iff the element A is the result of applying a finite sequence of the operations "$\cup$", "$\cap$", or "$^{-}$" to (some of the) elements $A_1, A_2, \ldots, A_n$. In this section we discuss two special forms in which an element may be expressed in terms of $A_1, A_2, \ldots, A_n$. The main result of the section is that if an element is expressible in any way in terms of $A_1, A_2, \ldots, A_n$, then it is expressible in each of these two special forms. The question of the uniqueness of these representations is also discussed.

We shall need some definitions. The terminology introduced in these definitions can be motivated by recalling that the operations "$\cup$" and "$\cap$" are sometimes referred to as disjunction and conjunction, respectively.

DEFINITION 2.1 An *elementary conjunction* based on $A_1, A_2, \ldots, A_n$ is an expression

$$B_1 \cap B_2 \cap \ldots \cap B_n$$

where, for each i $(1 \leq i \leq n)$, B_i is either A_i or $\bar{A}_i$.

Notice that not every conjunction of the elements A_i is an elementary conjunction. An elementary conjunction based on $A_1, A_2, \ldots, A_n$ is a conjunction of exactly n elements; the first of these elements is either A_1 or $\bar{A}_1$; the second is either A_2 or $\bar{A}_2$; and so forth. Moreover, we are concerned here with the form in which an element is expressed, and not with the element itself. For example, $A_1 \cap \bar{A}_2$ is an elementary conjuction based on A_1, A_2, but if we choose to set $C = A_1 \cap \bar{A}_2$, then, of course, C is the same element as $A_1 \cap \bar{A}_2$, but the expression "C" is not an elementary conjunction based on A_1, A_2 since the form in which it is written does not satisfy the requirement of Definition 2.1. We shall usually omit the quotation marks around the name of an expression; for example, we shall denote both the element C and the expression "C" simply as C. The context will make clear which meaning is intended.

DEFINITION 2.2 A *disjunctive normal form* based on $A_1, A_2, \ldots, A_n$ is a disjunction of a finite number of distinct expressions each of which is an elementary conjunction based on $A_1, A_2, \ldots, A_n$.

The two elementary conjunctions $B_1 \cap B_2 \cap \ldots \cap B_n$ and $C_1 \cap C_2 \cap \ldots \cap C_n$ are *distinct* iff, for at least one value of i, $B_i \neq C_i$. Note that it is possible for two distinct elementary conjunctions to represent the same element of $\mathfrak{B}$ (Example 2.3c).

Here, again, our interest is in the form in which an element is written, and not in the element itself.

Example 2.3 (a) The expression

$$(A \cap B) \cup (\bar{A} \cap \bar{B})$$

is in disjunctive normal form based on A, B. It is also in disjunctive normal form based on $\bar{A}, B$, etc.

(b) The expression $(A \cap G) \cup \bar{A}$ is not in disjunctive normal form based on A, G. The same element may be written as

$$(A \cap G) \cup (\bar{A} \cap G)$$

and here it is in disjunctive normal form based on A, G.

(c) The expressions $A \cap \bar{G}$ and $\bar{A} \cap \bar{G}$ are two distinct elementary conjunctions based on A, G; yet $A \cap \bar{G} = L = \bar{A} \cap \bar{G}$.

(d) Consider the empty subset $\emptyset$ of $\mathfrak{B}$; the expression

$$\cup \emptyset$$

is in disjunctive normal form based on any set of elements of $\mathfrak{B}$ since every element of $\emptyset$ (there are none!) is an elementary conjunction based on these elements. Of course, every element of $\mathfrak{B}$ is an upper bound for the set $\emptyset$. Thus $\cup \emptyset = L$, and we conclude that L can be expressed in disjunctive normal form based on any set of elements.

THEOREM 2.4 If A is any element of a Boolean algebra $\mathcal{B}$ and if A is expressible in terms of the elements $A_1, A_2, \ldots, A_n$, then A can be written in disjunctive normal form based on $A_1, A_2, \ldots, A_n$.

Proof. Consider any expression giving A in terms of $A_1, A_2, \ldots, A_n$. By successive use of DeMorgan's laws and the equation $\bar{\bar{B}} = B$, we can express A in a form in which the operation "$\bar{\ }$" is applied only to the individual elements $A_1, A_2, \ldots, A_n$. That is, we can arrange that the operation "$\bar{\ }$" does not follow any application of the operations "$\cup$" or "$\cap$", nor does it follow any application of itself. Thus, we may confine our attention to expressions formed by application of the two operations "$\cup$" and "$\cap$" to the elements $A_1, A_2, \ldots, A_n, \bar{A}_1, \bar{A}_2, \ldots, \bar{A}_n$. The proof proceeds by induction on the total number of occurrences of these elements in the particular expression. If there is only one such occurrence, the expression must be either A_i or $\bar{A}_i$. It is left as an exercise (Problem 3) to show that each of these elements can be written in disjunctive normal form based on $A_1, A_2, \ldots, A_n$.

Next, consider an element E represented by an expression $\mathcal{E}$ in which the elements $A_1, A_2, \ldots, A_n, \bar{A}_1, \bar{A}_2, \ldots, \bar{A}_n$ occur a total of m times ($m > 1$), and assume that any expression with fewer than m occurrences can be put in disjunctive normal form. The expression $\mathcal{E}$ must be in one of the two forms

$$\mathcal{P} \cup \mathcal{Q} \quad \text{or} \quad \mathcal{P} \cap \mathcal{Q}$$

where each of $\mathcal{P}$ and $\mathcal{Q}$ is an expression with fewer than m occurrences of the elements $A_1, A_2, \ldots, A_n, \bar{A}_1, \bar{A}_2, \ldots, \bar{A}_n$. We consider these two cases separately.

If $\mathcal{E}$ is in the form $\mathcal{P} \cup \mathcal{Q}$, then, by the induction hypothesis, each of the expressions $\mathcal{P}$ and $\mathcal{Q}$ can be written in disjunctive normal form and, when this is done, the new expression for $\mathcal{P} \cup \mathcal{Q}$, with repetitions of elementary conjunctions deleted, expresses E in disjunctive normal form.

If $\mathcal{E}$ is in the form $\mathcal{P} \cap \mathcal{Q}$, then, again, each of $\mathcal{P}$ and $\mathcal{Q}$ can be written in disjunctive normal form and, by the distributive law, E can be written as a disjunction of terms each one of which is a conjunction of an elementary conjunction from $\mathcal{P}$ with an elementary conjunction from $\mathcal{Q}$ (Problem 4). A typical one of these terms would be the conjunction

$$T = (B_1 \cap B_2 \cap \ldots \cap B_n) \cap (C_1 \cap C_2 \cap \ldots \cap C_n)$$

where, for each i ($1 \leq i \leq n$), B_i is either A_i or $\bar{A}_i$, and C_i is either A_i or $\bar{A}_i$. If, for each i ($1 \leq i \leq n$), $C_i = B_i$, then T is equal to the elementary conjunction $B_1 \cap B_2 \cap \ldots \cap B_n$. If, for at least one i, $C_i \neq B_i$, then $T = L$. Thus E can be written as a disjunction of terms each one of which is either the element L or is an elementary conjunction based on $A_1, A_2, \ldots,$

A_n. The occurrences of the element L may be simply dropped from this disjunction without changing the element which it represents. Thus E can be written as a disjunction of elementary conjunctions based on $A_1, A_2, \ldots, A_n$. ∎

The proof of Theorem 2.4 actually supplies a constructive procedure for writing an element A in disjunctive normal form if some expression representing A is given. The algebraic manipulations can usually be performed a little more simply by a different procedure, as illustrated in the following example.

Example 2.5 Write the element

$$\overline{[A \cap (\bar{B} \cup C)]} \cap \overline{[\bar{B} \cup \overline{(A \cup \bar{C})}]}$$

in disjunctive normal form based on A, B, C.

STEP 1. Use DeMorgan's laws and the equation $\bar{\bar{D}} = D$ to arrange that the operation "$^{-}$" is applied only to A, B, or C.

$$\overline{[A \cap (\bar{B} \cup C)]} \cap \overline{[\bar{B} \cup \overline{(A \cup \bar{C})}]} = [\bar{A} \cup (B \cap \bar{C})] \cap [B \cap (A \cup \bar{C})].$$

STEP 2. Use the distributive law to write the result of step 1 as a disjunction in which each of the terms is a conjunction of some of the elements $A, B, C, \bar{A}, \bar{B}, \bar{C}$. Simplify by eliminating repetitions and dropping the element L from the disjunction.

$$\begin{aligned}
[\bar{A} \cup (B \cap \bar{C})] \cap [B \cap (A \cup \bar{C})] \\
&= [\bar{A} \cup (B \cap \bar{C})] \cap [(B \cap A) \cup (B \cap \bar{C})] \\
&= (\bar{A} \cap B \cap A) \cup (\bar{A} \cap B \cap \bar{C}) \cup (B \cap \bar{C} \cap B \cap A) \cup (B \cap \bar{C} \cap B \cap \bar{C}) \\
&= L \cup (\bar{A} \cap B \cap \bar{C}) \cup (A \cap B \cap \bar{C}) \cup (B \cap \bar{C}) \\
&= (\bar{A} \cap B \cap \bar{C}) \cup (A \cap B \cap \bar{C}) \cup (B \cap \bar{C}).
\end{aligned}$$

STEP 3. For each term of the result of step 2 which is not already an elementary conjunction based on A, B, C, write that term as the inf of itself and G with G written in a form suitable for that particular term. Again use the distributive law and simplify. In the present example, $B \cap \bar{C}$ is the only term which is not already an elementary conjunction based on A, B, C. Since we require an occurrence of A or $\bar{A}$ in this term, we write it as $[(A \cup \bar{A}) \cap (B \cap \bar{C})]$ obtaining

$$\begin{aligned}
(\bar{A} \cap B \cap \bar{C}) \cup (A \cap B \cap \bar{C}) \cup [(A \cup \bar{A}) \cap (B \cap \bar{C})] \\
&= (\bar{A} \cap B \cap \bar{C}) \cup (A \cap B \cap \bar{C}) \cup (A \cap B \cap \bar{C}) \cup (\bar{A} \cap B \cap \bar{C}) \\
&= (\bar{A} \cap B \cap \bar{C}) \cup (A \cap B \cap \bar{C}).
\end{aligned}$$

We turn now to the question of uniqueness of the disjunctive normal form for a particular element. We have seen that, for each element A of $\mathcal{B}$ which is expressible in terms of $A_1, A_2, \ldots, A_n$, there is a set $\mathfrak{S}$ of elementary conjunctions based on $A_1, A_2, \ldots, A_n$ such that

$$A = \bigcup \mathfrak{S}.$$

We inquire whether or not the set $\mathfrak{S}$ is uniquely determined by A and $A_1, A_2, \ldots, A_n$. A simple necessary and sufficient condition is given in the next theorem.

THEOREM 2.6 Let A be expressible in terms of $A_1, A_2, \ldots, A_n$. The disjunctive normal form for A based on $A_1, A_2, \ldots, A_n$ is unique if and only if the element L cannot be written as an elementary conjunction based on $A_1, A_2, \ldots A_n$.

Proof. It is easy to see that if L can be written as an elementary conjunction based on $A_1, A_2, \ldots, A_n$, then no disjunctive normal form based on $A_1, A_2, \ldots, A_n$ can be unique. In fact, if $\mathfrak{S}$ is any set of elementary conjunctions based on $A_1, A_2, \ldots, A_n$, the elementary conjunction representing L could be either included in $\mathfrak{S}$ or excluded from it without changing the element represented by the expression $\bigcup \mathfrak{S}$.

Conversely, suppose there are two different disjunctive normal forms based on $A_1, A_2, \ldots, A_n$ which represent the same element. Say

$$\bigcup \mathfrak{S}_1 = \bigcup \mathfrak{S}_2$$

with a particular elementary conjunction $B_1 \cap B_2 \cap \ldots \cap B_n$ in $\mathfrak{S}_1$ but not in $\mathfrak{S}_2$. Then

$$(B_1 \cap B_2 \cap \ldots \cap B_n) \cap (\bigcup \mathfrak{S}_1) = (B_1 \cap B_2 \cap \ldots \cap B_n) \cap (\bigcup \mathfrak{S}_2).$$

Using the distributive law, and noting that two elementary conjunctions must either be identical or else their inf must be L (Problem 5), the last equation reduces to

$$B_1 \cap B_2 \cap \ldots \cap B_n = L. \blacksquare$$

An interesting and important extension of these results is given in Problems 9 and 10. We conclude this section by stating a sequence of definitions and theorems which are analogous to the results we have developed.

DEFINITION 2.7 An *elementary disjunction* based on $A_1, A_2, \ldots, A_n$ is an expression

$$B_1 \cup B_2 \cup \ldots \cup B_n$$

where, for each i $(1 \leq i \leq n)$, B_i is either A_i or $\bar{A}_i$.

DEFINITION 2.8 A *conjunctive normal form* based on $A_1, A_2, \ldots, A_n$ is a conjunction of a finite number of distinct expressions each of which is an elementary disjunction based on $A_1, A_2, \ldots, A_n$.

THEOREM 2.9 If A is any element of a Boolean algebra $\mathcal{B}$ and if A is expressible in terms of the elements $A_1, A_2, \ldots, A_n$, then A can be written in conjunctive normal form based on $A_1, A_2, \ldots, A_n$.

Proof. Problem 7a. ∎

THEOREM 2.10 Let A be expressible in terms of $A_1, A_2, \ldots, A_n$. The conjunctive normal form for A based on $A_1, A_2, \ldots, A_n$ is unique if and only if the element G cannot be written as an elementary disjunction based on $A_1, A_2, \ldots, A_n$.

Proof. Problem 7b. ∎

PROBLEMS

1. Find a disjunctive normal form based on A, B, C, for each of the following.

   (a) A.

   (b) $\overline{\overline{\bar{A} \cup B} \cup C}$.

   (c) $A \cup \{\overline{B \cap [C \cup (\overline{A \cap B})]}\}$.

2. Find a conjunctive normal form based on A, B, C, for each of the elements in Problem 1.

3. (a) Prove that, for each i $(1 \leq i \leq n)$, A_i can be written in disjunctive normal form based on $A_1, A_2, \ldots, A_n$.

   (b) Prove also that $\bar{A}_i$ can be written in disjunctive normal form based on $A_1, A_2, \ldots, A_n$.

4. Let $\mathcal{P}$ and $\mathcal{Q}$ be two expressions in disjunctive normal form based on $A_1, A_2, \ldots, A_n$. Prove that the element $\mathcal{P} \cap \mathcal{Q}$ is the disjunction of

the set of all elements which can be written as a conjunction of an elementary conjunction of $\mathcal{P}$ with an elementary conjunction of $\mathcal{Q}$.

5. Prove that two elementary conjunctions based on $A_1, A_2, \ldots, A_n$ must be identical or their inf must be L.

6. Let A and B be expressible in terms of $A_1, A_2, \ldots, A_n$. Prove that if the disjunctive normal form for A based on $A_1, A_2, \ldots, A_n$ is unique, then that for B is also unique.

7. (a) Prove Theorem 2.9.

(b) Prove Theorem 2.10.

8. Would an exhaustive table of disjunctive normal forms be of any help in finding a conjunctive normal form?

#9. Let X be a Boolean algebra containing elements $A_1, A_2, \ldots, A_n$ such that each element of $\mathcal{B}$ has a unique disjunctive normal form based on $A_1, A_2, \ldots, A_n$. Prove each of the theorems (a) through (c) below.

(a) Each atom of $\mathcal{B}$ is an elementary conjunction based on $A_1, A_2, \ldots, A_n$.

(b) Each elementary conjunction based on $A_1, A_2, \ldots, A_n$ is an atom of $\mathcal{B}$. (*Hint:* Use Lemma 3.4, section 5-3, and Problem 5.)

(c) $\mathcal{B}$ is isomorphic to the Boolean algebra $\mathcal{B}_n$ of all Boolean functions of n independent variables. (*Hint:* Use Problem 8, Section 5-3.)

These results show that if there exist elements $A_1, A_2, \ldots, A_n$ in $\mathcal{B}$ such that disjunctive normal forms based on $A_1, A_2, \ldots, A_n$ are unique, then $\mathcal{B}$ is isomorphic to $\mathcal{B}_n$. We shall see in Section 7-4 that the converse of this result is also correct.

#10. Let $A_1, A_2, \ldots, A_n$ be elements of a Boolean algebra $\mathcal{B}$. A disjunctive normal form for A, based on $A_1, A_2, \ldots, A_n$ is said to be *minimal* iff it involves fewer elementary conjunctions than any other disjunctive normal form for A, based on $A_1, A_2, \ldots, A_n$. Prove that if A is expressible in terms of $A_1, A_2, \ldots, A_n$, then there is a unique minimal disjunctive normal form for A based on $A_1, A_2, \ldots, A_n$. (*Hint:* Use the ideas in the proof of Theorem 2.6.)

11. (a) Let a, b, and c be three distinct objects, and let $\mathcal{B}$ be the Boolean algebra of all subsets of the set $S = \{a, b, c\}$. Prove directly that if $A_1, A_2, \ldots, A_n$ are elements of $\mathcal{B}$ such that every element of $\mathcal{B}$ can be written in disjunctive normal form based on $A_1, A_2, \ldots, A_n$, then these disjunctive normal forms are not unique.

(b) Let $\mathcal{B}$ be the Boolean algebra of all ordered quintuples of Boolean constants. Set $A_1 = (1, 0, 1, 0, 1)$, $A_2 = (1, 0, 0, 1, 0)$, $A_3 = (1, 1, 0, 0, 1)$. Find all the disjunctive normal forms for the element $A = (0, 1, 0, 1, 0)$ based on A_1, A_2, A_3.

7-3 The Ring Normal Form

In Section 7-2 we have developed standard forms for expressing an element of a Boolean algebra using the operations "$\cup$", "$\cap$", and "$\bar{\ }$" of the Boolean algebra. In this section we consider a standard form using the ring operations "$+$" and "$\cdot$". Let $\mathcal{R}$ be a Boolean ring with a unit element; that is, $\mathcal{R}$ is a Boolean ring associated with some Boolean algebra as discussed in Section 6-3. Intuitively, the ring standard form is obtained by "multiplying out and simplifying" using the special rules for algebra in a Boolean ring, but we shall need a more explicit definition.

DEFINITION 3.1 An *elementary monomial* based on the elements $1, A_1, A_2, \ldots, A_n$ of a Boolean ring $\mathcal{R}$ is either 1 or an expression

$$A_{i_1}A_{i_2}\ldots A_{i_r} \quad \text{where} \quad i_1 < i_2 < \ldots < i_r.$$

DEFINITION 3.2 An *elementary monomial* based on the elements $A_1, A_2, \ldots, A_n$ of a Boolean ring $\mathcal{R}$ with a unit element (none of the A_i is the unit element) is an expression

$$A_{i_1}A_{i_2}\ldots A_{i_r} \quad \text{where} \quad i_1 < i_2 < \ldots < i_r.$$

The more important case is that of Definition 3.1 where the unit element is included among the base elements. The subsequent results in this section will be stated for this case only; the case where the unit element is not included in the base elements is considered in Problems 7 and 8.

DEFINITION 3.3 A *ring normal form* based on $1, A_1, A_2, \ldots, A_n$ is either 0 or a sum of distinct elementary monomials based on $1, A_1, A_2, \ldots, A_n$.

Just as in the discussion of the disjunctive normal form in Section 7-2, we are concerned with the particular form in which an expression is written and not merely with the element of $\mathcal{R}$ which that expression represents. Two elementary monomials

$$A_{i_1}A_{i_2}\ldots A_{i_r} \quad \text{and} \quad A_{j_1}A_{j_2}\ldots A_{j_s}$$

are *distinct* iff either $r \neq s$ or there is a k such that $A_{i_k} \neq A_{j_k}$.

Example 3.4 Each of the following is in ring normal form based on 1, A, B, C.

(a) $A + B$.

(b) $B + A$.

(c) $1 + A + ABC$.

(d) $A + AB + ABC$.

Example 3.5 None of the following is in ring normal form based on 1, A, B, C.

(a) $B + 0$.

(b) $A + AB + A$.

(c) $C + ABA$.

(d) BA.

It is natural to inquire which elements of $\mathcal{R}$ can be expressed in ring normal form based on a particular set of elements, and under what circumstances each element of $\mathcal{R}$ determines a unique ring normal form. The first of these questions is answered in Theorem 3.6. A partial answer to the second question is given in Theorem 3.8; a complete answer will be found in Section 7-4. We shall say that an element A of $\mathcal{R}$ is *expressible* in terms of $1, A_1, A_2, \ldots, A_n$ iff the element is the result of applying a finite sequence of the operations "+" or "·" to (some of) the elements $1, A_1, A_2, \ldots, A_n$.

THEOREM 3.6 If A is expressible in terms of $1, A_1, \ldots, A_n$, then A can be written in ring normal form based on $1, A_1, A_2, \ldots, A_n$.

Proof. Having given an expression which represents the element A and which involves only the operations "+" and "·" and the elements 1, $A_1, A_2, \ldots, A_n$, the distributive law enables us to write A as a sum of monomials (but not necessarily elementary monomials). The commutative and associative laws for multiplication, together with the laws

$$B^2 = B \quad \text{and} \quad 1 \cdot B = B,$$

allow us to reduce each of these monomials to an elementary monomial. Finally, the commutative and associative laws for addition, together with the laws

$$B + B = 0 \quad \text{and} \quad B + 0 = B,$$

enable us to eliminate repetitions among the elementary monomials, thus expressing A in ring normal form based on $1, A_1, A_2, \ldots, A_n$. ∎

The steps in the proof of Theorem 3.6 are illustrated in Example 3.7. The algebraic manipulations required to put a given expression in ring normal form are frequently simpler than those required in the reduction to disjunctive normal form. This simplicity is one of the main advantages of the ring normal form. The reader should compare the work of Example 3.7 with that of Example 2.5.

Example 3.7 Write the element

$$(A + B)\,(1 + C)AB + (A + B + C)\,(1 + A)$$

in ring normal form based on $1, A, B, C$.

STEP 1. Applying the distributive law, the element may be written as

$$A1AB + ACAB + B1AB + BCAB + A1 + B1 + C1 + A^2 + BA + CA.$$

STEP 2. The commutative and associative laws for multiplication, together with the laws $D^2 = D$ and $1 \cdot D = D$, enable us to reduce this expression to

$$AB + ABC + AB + ABC + A + B + C + A + AB + AC.$$

STEP 3. Finally, application of the commutative and associative laws for addition, together with the laws $D + D = 0$ and $D + 0 = D$, gives the following ring normal form.

$$B + C + AB + AC.$$

We come now to the question of uniqueness of the ring normal form. Of course, the order in which the elementary monomials are written in a ring normal form is irrelevant; we shall say that the ring normal form for A based on $1, A_1, A_2, \ldots, A_n$ is unique iff the element A of $\mathcal{R}$ determines a unique collection of elementary monomials, based on $1, A_1, A_2, \ldots, A_n$, whose sum is A.

THEOREM 3.8 If every element A of $\mathcal{R}$ has a unique ring normal form based on $1, A_1, A_2, \ldots, A_n$, then $\mathcal{R}$ has exactly 2^{2^n} elements.

Proof. The proof consists in counting the number of elementary monomials based on $1, A_1, A_2, \ldots, A_n$, then counting the number of sets of these elementary monomials. Since each element of $\mathcal{R}$ has a unique ring normal form, the number of elements in $\mathcal{R}$ is the same as the number of sets of elementary monomials.

COUNTING THE ELEMENTARY MONOMIALS: The expression 1 is an elementary monomial. Any other elementary monomial is a product

$$A_{i_1}A_{i_2}\ldots A_{i_r} \quad \text{where} \quad 1 \leq i_1 < i_2 < \ldots < i_r \leq n.$$

If we remove the inequality restrictions between the subscripts, there are $r!$ different orders in which a particular set of r different factors can be written. Also, every product of r different factors from $A_1, A_2, \ldots, A_n$ can be arranged in exactly one order so that the inequality restrictions between the subscripts are satisfied. Thus, if we count the number of products of r different factors and divide by $r!$, we shall obtain the number of products of r factors with the inequalities satisfied. For r different factors, the first factor may be chosen in n ways, the second in $n - 1$ ways, and so on; finally, the last (r^{th}) factor can be chosen in $n - r + 1$ ways. Thus, the number of products of r different factors is $n(n-1)\ldots(n-r+1)$, and the number of elementary monomials of the form $A_{i_1}A_{i_2}\ldots A_{i_r}$ is

$$\frac{n(n-1)(n-2)\ldots(n-r+1)}{r!}.$$

The total number of elementary monomials is 1 (for the expression "1") plus the number of elementary monomials of the form A_i, plus the number of elementary monomials of the form $A_{i_1}A_{i_2}$, etc. That is,

$$1 + \frac{n}{1} + \frac{n(n-1)}{2!} + \frac{n(n-1)(n-2)}{3!} + \ldots + \frac{n!}{n!}.$$

By Problem 6, this sum is 2^n.

COUNTING THE SETS OF ELEMENTARY MONOMIALS: The set S of *all* elementary monomials has 2^n elements; we want to count the subsets of S. We can choose a subset of S by deciding, for each element of S, whether to include that element or exclude it. Since, for each element of S, we have a choice between two alternatives, and there are 2^n elements of S, the total number of ways in which these choices can be made is 2^{2^n}. ∎

We shall see in the next section that, if $\mathcal{R}$ has 2^{2^n} elements, then it is possible to choose elements $1, A_1, A_2, \ldots, A_n$ of $\mathcal{R}$ so that each element A of $\mathcal{R}$ has a unique ring normal form based on $1, A_1, A_2, \ldots, A_n$.

PROBLEMS

1. Express each of the following elements in ring normal form based on the elements $1, A, B, C$ of a Boolean ring $\mathcal{R}$.

   (a) $A(A + B)(A + B + C)$.

(b) $(1 + A)(1 + B)(1 + C)$.

(c) $[(1 + A) + (1 + A + B)](1 + A + B + C)$.

(d) $(1 + A)(1 + AB)(1 + ABC)(A + B + C)$.

2. (a) Prove that if the element A has a unique ring normal form based on $1, A_1, A_2, \ldots, A_n$, then the elements $1, A_1, A_2, \ldots, A_n$ are different.

(b) Give an example of a Boolean ring $\mathcal{R}$ containing $n + 1$ different elements $1, A_1, A_2, \ldots, A_n$ such that there is an element A of $\mathcal{R}$ with two different ring normal forms based on $1, A_1, A_2, \ldots, A_n$.

3. Show that, for any elements $1, A_1, A_2, \ldots, A_n$ of $\mathcal{R}$, the element 0 is expressible in terms of $1, A_1, A_2, \ldots, A_n$.

#4. Show that a necessary and sufficient condition that an element A of $\mathcal{R}$ has a unique ring normal form based on $1, A_1, A_2, \ldots, A_n$ is that A is expressible in terms of $1, A_1, A_2, \ldots, A_n$, and the expression "0" is the only ring normal form for the element 0, based on $1, A_1, A_2, \ldots, A_n$.

#5. Let $\mathcal{B}$ and $\mathcal{R}$ be, respectively, a Boolean algebra and the Boolean ring associated with it as in Section 6-3, and let $A_1, A_2, \ldots, A_n$ be elements of $\mathcal{B}$ (and $\mathcal{R}$). Prove that A is expressible (by "$\cup$", "$\cap$", and "$\overline{\ }$") in terms of $A_1, A_2, \ldots, A_n$ if and only if A is expressible (by "+" and "$\cdot$") in terms of $1, A_1, A_2, \ldots, A_n$.

6. Prove that

$$1 + \frac{n}{1} + \frac{n(n-1)}{2!} + \frac{n(n-1)(n-2)}{3!} + \ldots + \frac{n!}{n!} = 2^n.$$

[*Hint:* Expand $(1 + 1)^n$ by the binomial theorem.]

7. Extend the definitions and theorems of this section to ring normal forms based on elements $A_1, A_2, \ldots, A_n$ different from the unit element. (*Hint:* The analog of Theorem 3.8 is as follows: If every element A of $\mathcal{R}$ has a unique ring normal form based on $A_1, A_2, \ldots, A_n$, then $\mathcal{R}$ has exactly 2^{2^n-1} elements.)

8. (a) Consider the Boolean ring $\mathcal{R}$ of all ordered triples of Boolean constants. Find n elements $A_1, A_2, \ldots, A_n$ of $\mathcal{R}$, such that every element A of $\mathcal{R}$ has a unique ring normal form based on $A_1, A_2, \ldots, A_n$.

(b) In part (a), is the set $\{A_1, A_2, \ldots, A_n\}$ unique? Is the integer n unique?

9. In the Boolean ring $\mathcal{R}$ of all ordered triples of Boolean constants, show, by direct computation, that there is no set of elements $A_1, A_2, \ldots, A_n$ such that every element A of $\mathcal{R}$ has a unique ring normal form based on $1, A_1, A_2, \ldots, A_n$.

ξ_1	ξ_2	ξ_3	ψ_1	ψ_2	ψ_3	ψ_4
1	1	1	1	1	1	1
1	1	0	1	0	1	0
1	0	1	0	0	1	0
1	0	0	0	0	1	1
0	1	1	0	0	1	0
0	1	0	0	1	1	0
0	0	1	1	0	0	0
0	0	0	1	0	0	1

FIGURE 4.1

10. (**a**) In the Boolean ring $\mathcal{R}$ of all ordered pairs of Boolean constants, find a set of elements $A_1, A_2, \ldots, A_n$ such that every element A of $\mathcal{R}$ has a unique ring normal form based on $1, A_1, A_2, \ldots, A_n$.

(**b**) Do part (a) for the ring of all ordered quadruples of Boolean constants.

#11. Prove that if every element A of a Boolean ring $\mathcal{R}$ has a unique ring normal form based on $1, A_1, A_2, \ldots, A_n$, then $\mathcal{R}$ is isomorphic to the ring obtained from the Boolean algebra $\mathcal{B}_n$, of all Boolean functions of n independent variables, by the process of Theorem 3.1, Section 6-3.

7-4 Normal Forms in $\mathcal{B}_n$

We have seen (Problem 9, Section 7-2 and Problem 11, Section 7-3) that the only cases in which normal forms might be unique involve the Boolean algebra $\mathcal{B}_n$ of all Boolean functions of n independent variables. In this section we consider this Boolean algebra $\mathcal{B}_n$ and the Boolean ring $\mathcal{R}$ obtained from $\mathcal{B}_n$ as in Theorem 3.1 of Section 6-3. We show that there are n elements $\phi_1, \phi_2, \ldots, \phi_n$ of $\mathcal{B}_n$ such that every element ϕ of $\mathcal{B}_n$ has a unique disjunctive normal form based on $\phi_1, \phi_2, \ldots, \phi_n$. Moreover, every element ϕ of $\mathcal{R}$ has a unique ring normal form based on $1, \phi_1, \phi_2, \ldots, \phi_n$. (Here "1" denotes the Boolean function which is identically equal to 1.)

The set $\{\phi_1, \phi_2, \ldots, \phi_n\}$ of Boolean functions is not unique (Problem 3), but we shall find it convenient to use the functions defined by

$$(*) \qquad \phi_i(\alpha_1, \alpha_2, \ldots, \alpha_n) = \alpha_i. \qquad (i = 1, 2, \ldots, n)$$

Figure 4.1 shows the tabular representations of these functions, and of the function 1, for the special case of three independent variables ξ_1, ξ_2, ξ_3.

THEOREM 4.1 Every element ϕ of $\mathcal{B}_n$ has a unique disjunctive normal form based on the functions $\phi_1, \phi_2, \ldots, \phi_n$ defined by Equation (*).

Proof. By Lemmas 3.5 and 3.6, Section 5-3, any element of a Boolean algebra is uniquely expressible as a sup of atoms. The atoms of $\mathcal{B}_n$ are the Boolean functions which take on the value 1 exactly once. The proof will be completed by showing that these atoms in $\mathcal{B}_n$ are uniquely expressible as elementary conjunctions based on $\phi_1, \phi_2, \ldots, \phi_n$. In fact, these elementary conjunctions are of the form

$$\psi_1 \cap \psi_2 \cap \ldots \cap \psi_n$$

where, for each i $(1 \leq i \leq n)$ ψ_i is either ϕ_1 or $\bar{\phi}_i$. Any given atom in $\mathcal{B}_n$ determines a unique row in the tabular form where the atom takes on the value 1. This row determines a unique set of values $(\alpha_1, \alpha_2, \ldots, \alpha_n)$ for the n independent variables, and this set of values determines a unique elementary conjunction by setting

$$\psi_i = \begin{cases} \phi_i, & \text{if} \quad \alpha_i = 1 \\ \bar{\phi}_i, & \text{if} \quad \alpha_i = 0. \end{cases}$$

It is left as an exercise to show that

$$\psi_1 \cap \psi_2 \cap \ldots \cap \psi_n$$

expresses the given atom, and that this is the only elementary conjunction which does so (Problem 4). ∎

Example 4.2 Find the disjunctive normal form, based on ϕ_1, ϕ_2, ϕ_3, for the function ϕ whose table of values is shown in Fig. 4.2. The elementary

FIGURE 4.2

ξ_1	ξ_2	ξ_3	ϕ
1	1	1	1
1	1	0	0
1	0	1	0
1	0	0	1
0	1	1	0
0	1	0	0
0	0	1	1
0	0	0	0

conjunction $\phi_1 \cap \phi_2 \cap \phi_3$ represents the atom with its "1" in the first row. Similarly, $\phi_1 \cap \bar{\phi}_2 \cap \bar{\phi}_3$ and $\bar{\phi}_1 \cap \bar{\phi}_2 \cap \phi_3$ represent the atoms with a "1" in the 4th and 7th rows, respectively. Thus,

$$\phi = (\phi_1 \cap \phi_2 \cap \phi_3) \cup (\phi_1 \cap \bar{\phi}_2 \cap \bar{\phi}_3) \cup (\bar{\phi}_1 \cap \bar{\phi}_2 \cap \phi_3).$$

We turn now to the ring $\mathcal{R}$ obtained from $\mathcal{B}_n$. We shall continue to use the notation ϕ_i for the special functions defined by Equation (*) and we shall be interested in the ring operations "+" and "·" as well as in the Boolean algebra operations "$\cup$", "$\cap$", and "$^-$".

THEOREM 4.3 Every element ϕ of $\mathcal{R}$ has a unique ring normal form based on the functions $1, \phi_1, \phi_2, \ldots, \phi_n$.

Proof. From Theorem 4.1, every element ϕ of $\mathcal{R}$ is expressible (by the operations "$\cup$", "$\cap$", and "$^-$") in terms of $\phi_1, \phi_2, \ldots, \phi_n$. By Problem 5, Section 7-3, ϕ is expressible (by the operations "+" and "·") in terms of 1, $\phi_1, \phi_2, \ldots, \phi_n$. It follows from Theorem 3.6, Section 7-3, that ϕ can be expressed in ring normal form based on $1, \phi_1, \phi_2, \ldots, \phi_n$.

It remains to show that the ring normal form is unique. By Problem 4, Section 7-3, it suffices to show that the expression "0" is the only ring normal form for the element 0. Suppose there is a different ring normal form for 0. From the elementary monomials in this normal form, choose one, say

$$\phi_{i_1} \phi_{i_2} \cdots \phi_{i_r}$$

which involves a minimum number of the functions $\phi_1, \phi_2, \ldots, \phi_n$. For the set $(\alpha_1, \alpha_2, \ldots, \alpha_n)$ of Boolean constants defined by

$$\alpha_i = \begin{cases} 1, & \text{for} \quad i = i_1, i_2, \ldots, i_r \\ 0, & \text{otherwise,} \end{cases}$$

the monomial $\phi_{i_1}, \phi_{i_2}, \ldots \phi_{i_r}$, has the value one, and each of the other elementary monomials in this normal form has the value zero. Thus this normal form cannot represent the function 0, since its value at $(\alpha_1, \alpha_2, \ldots, \alpha_n)$ is one. ∎

PROBLEMS

1. Four Boolean functions, $\psi_1, \psi_2, \psi_3, \psi_4$, each of three independent variables, are defined by the table in Fig. 4.3. Express each of these functions in disjunctive normal form based on the functions ϕ_1, ϕ_2, ϕ_3 of this section.

ξ_1	ξ_2	ξ_3	1	ϕ_1	ϕ_2	ϕ_3
1	1	1	1	1	1	1
1	1	0	1	1	1	0
1	0	1	1	1	0	1
1	0	0	1	1	0	0
0	1	1	1	0	1	1
0	1	0	1	0	1	0
0	0	1	1	0	0	1
0	0	0	1	0	0	0

FIGURE 4.3

2. (a) Express each of the functions $\psi_1, \psi_2, \psi_3, \psi_4$, of Problem 1, in ring normal form based on $1, \phi_1, \phi_2, \phi_3$.

   (b) Express each of the functions $\psi_1, \psi_2, \psi_3, \psi_4$, of Problem 1, in conjunctive normal form based on ϕ_1, ϕ_2, ϕ_3.

3. Find a set of Boolean functions $\{\psi_1, \psi_2, \ldots, \psi_n\}$, different from the set $\{\phi_1, \phi_2, \ldots, \phi_n\}$ such that every element of $\mathcal{B}_n$ has a unique disjunctive normal form based on $\psi_1, \psi_2, \ldots, \psi_n$.

4. Complete the proof of Theorem 4.1.

5. (a) Prove that the elementary monomial 1 cannot appear in a ring normal form for zero based on $1, \phi_1, \phi_2, \ldots, \phi_n$. Does the proof given for Theorem 4.3 cover this case?

   (b) Find a set of elements $\psi_1, \psi_2, \ldots, \psi_n$ of $\mathcal{B}_n$ such that the elementary monomial 1 does appear in a ring normal form for zero based on $1, \psi_1, \psi_2, \ldots, \psi_n$.

6. Prove that each element ϕ of $\mathcal{B}_n$ has a unique conjunctive normal form based on $\phi_1, \phi_2, \ldots, \phi_n$.

7-5 Duality

The principle of duality is usually introduced quite early in the study of Boolean algebras since it enables some proofs to be given very easily. For example, our results on conjunctive normal forms would follow immediately from those on disjunctive normal forms. We have purposely left the discussion of duality to the last, and have posed as exercises many of the theorems whose proofs could have been given by means of this principle. It is our hope that, in doing these exercises without the principle of duality, the student will have carefully thought through the proofs of the companion theorems in the text. In fact, the better students will certainly have discovered the principle of duality through these exercises. For this reason, we give only a brief discussion of it here.

The principle of duality is based on a certain symmetry in the axiom system defining a Boolean algebra. This symmetry can be seen in the characterization of a Boolean algebra given in Theorem 2.6, Section 5-2. Suppose we interchange the operations "$*$" and "$\circ$", and interchange the special elements P and Q in each of the conditions (a) through (f) of Theorem 2.6, Section 5-2. In this process, only condition (e) is changed and, by Theorem 3.6, Section 4-3, it is changed into a requirement which is equivalent to the original condition (in the presence of the other conditions). Thus, after the interchanges have been made, the conditions still characterize a Boolean algebra. In the usual notation, we have interchanged "$\cup$" with "$\cap$", and L with G.

Now consider any theorem in a Boolean algebra $\mathfrak{B}$. If we interchange "$\cup$" with "$\cap$" and L with G in the statement of this theorem, the result will still be a theorem, called the dual of the original theorem. A proof of this dual theorem can be obtained by making the interchanges in the proof of the original theorem, both in the steps of the proof and in the reasons for these steps. An illustration is given in Example 5.1 where the proofs of two dual theorems are presented.

Example 5.1 In parts (a) and (b), proofs are presented for two very simple dual theorems. Only the steps of the proofs are given; the reasons are to be supplied as an exercise (Problem 3).

(a) THEOREM: If A and B are elements of a Boolean algebra, then

$$A \cup (\bar{A} \cap B) \cup L = A \cup B.$$

PROOF:
$$\begin{aligned} A \cup (\bar{A} \cap B) \cup L &= A \cup (\bar{A} \cap B) \\ &= (A \cup \bar{A}) \cap (A \cup B) \\ &= G \cap (A \cup B) \\ &= A \cup B. \end{aligned}$$

(b) THEOREM: If A and B are elements of a Boolean algebra $\mathfrak{B}$, then

$$A \cap (\bar{A} \cup B) \cap G = A \cap B.$$

PROOF:
$$\begin{aligned} A \cap (\bar{A} \cup B) \cap G &= A \cap (\bar{A} \cup B) \\ &= (A \cap \bar{A}) \cup (A \cap B) \\ &= L \cup (A \cap B) \\ &= A \cap B. \end{aligned}$$

We shall see in the next chapter that the principle of duality is also of interest in connection with the applications of Boolean algebra. It allows us to choose between two formulations for many problems.

PROBLEMS

1. Use the duality principle to obtain our results on conjunctive normal forms from those on disjunctive normal forms.

2. Is there a principle of duality in a lattice? If so, what is it?

3. Supply the reasons for the steps in the proofs of Example 5.1. Are the reasons for corresponding steps dual to each other?

4. Prove that if an element A of a Boolean algebra can be expressed (by the operations "$\cup$", "$\cap$", and "$\bar{\ }$" in terms of the elements $A_1, A_2, \ldots, A_n$, then $\bar{A}$ can also be so expressed and, in fact, an expression for $\bar{A}$ can be obtained by interchanging "$\cup$" with "$\cap$" and A_i with $\bar{A}_i$ ($i = 1, 2, \ldots, n$).

5. Let $E(L, G, X_1, X_2, \ldots, X_n)$ and $F(L, G, X_1, X_2, \ldots, X_n)$ be two formal expressions formed from the symbols $L, G, X_1, X_2, \ldots, X_n$ by applying the operations "$\cup$", "$\cap$" and "$\bar{\ }$". Let $E'(G, L, X_1, X_2, \ldots, X_n)$ and $F'(G, L, X_1, X_2, \ldots, X_n)$ be the expressions obtained from E and F respectively by interchanging "$\cup$" with "$\cap$" and L with G. Now let L and G be, respectively, the least and greatest elements of a Boolean algebra $\mathfrak{B}$. Prove the following theorem directly, without using the duality principle.

   THEOREM: If

   $$E(L, G, B_1, B_2, \ldots, B_n) = F(L, G, B_1, B_2, \ldots, B_n)$$

   for all choices of elements $B_1, B_2, \ldots, B_n$ of $\mathfrak{B}$, then

   $$E'(G, L, B_1, B_2, \ldots, B_n) = F'(G, L, B_1, B_2, \ldots, B_n)$$

   is similarly valid. [*Hint:* Apply Problem 4 to the equation $E(L, G, \bar{B}_1, \bar{B}_2, \ldots, \bar{B}_n) = F(L, G, \bar{B}_1, \bar{B}_2, \ldots, \bar{B}_n)$.]

6. For any element A of a Boolean algebra, $A \cup \bar{A} = G$. What can be inferred from this by means of the principle of duality? By Problem 4? By Problem 5?

8

SOME APPLICATIONS OF BOOLEAN ALGEBRA

8-1 Introduction

Boolean algebra originated with the work of George Boole in the middle of the 19th century. He discussed the abstract rules governing the use of the logical connectives "$\wedge$", "$\vee$", "$\sim$", etc. For about a century, his work remained of interest mainly to logicians, but, in the middle of the 20th century, with the mushrooming growth of the computer industry, Boolean algebra became of importance to many scientists and engineers. The abstract mathematical system which Boole developed proved to be applicable to the design of computers and of electrical networks for various purposes. In Sections 8-2 and 8-3 we present an introduction to some of these applications. The mathematical ideas and procedures which we present are among those which are currently being used for design purposes, but the electrical hardware we mention is hopelessly inadequate for use in computers. We consider only the simplest of electrical hardware so that the discussion will be understandable to a student with no electrical background. Moreover, even if we were to present the latest advances in electrical hardware, the material would become obsolete in a very few years. The mathematical structure has proved to be more enduring. In Section 8-4 we consider some applications in the field of logic.

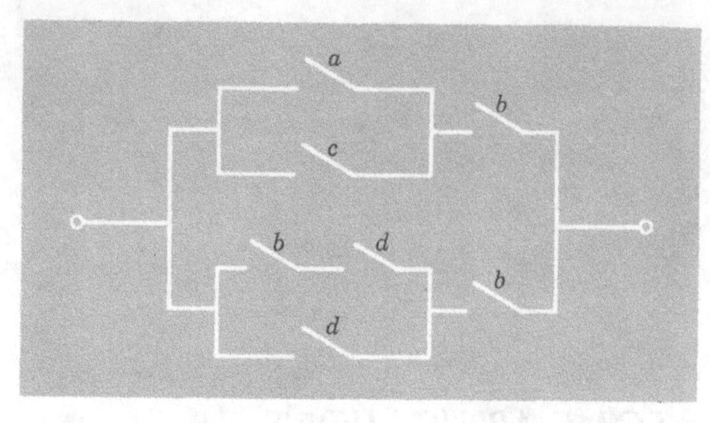

FIGURE 2.1

8-2 Applications to Electrical Networks

In this section we consider certain networks which have two terminals and which are composed of switches connected in various ways. An example is shown in Fig. 2.1.

A switch, as we shall consider it, is a very simple electrical device which has exactly two mutually exclusive states. A switch is either closed (allowing current to pass) or open (not allowing current to pass). We shall say that a closed switch is in *state one* and that an open switch is in *state zero.* In our figures, all switches will be depicted as being open. We shall use lower case letters as labels for switches, and shall use the corresponding capital letters as the states of the respective switches. Two switches may be coupled so that they are always in the same state; in this case, we use the same letter as the label for both of the switches as in Fig. 2.1. Two switches may be coupled so that they are always in different states; in this case, we shall label one of the switches a, for example, and the other one $\bar{a}$. The states of these two switches would be A and $\bar{A}$ respectively. We shall say that a network is in *state one* iff it allows current to pass between its terminals (i.e. iff the terminals are electrically connected); a network is in *state zero* iff it does not allow current to pass between its terminals (terminals not electrically connected). A visual indication of the state of a network can be obtained by closing the network by a return line connecting its two terminals and putting a light bulb and a battery in this return line. The light will be on if and only if the network is in state one.

We shall be interested in the following three types of problems: (1) Given a network, find all of the states of the respective switches so that the network is in state one. (2) Given a network, find, if possible, a simpler *equivalent* network; that is, find a simpler network which will be in the same state as the original network, no matter what states are assigned to the respective switches. (3) Given a number of switches, and a collection of assignments of states to these respective switches, find a network which will be in state one iff the assignment of states to the switches is one of the given ones. Examples 2.1 through 2.3 illustrate these three types of problems.

Two simple networks are shown in Fig. 2.2. Results concerning these simple networks will be useful in connection with the more complicated networks we shall consider. In Fig. 2.2a, the two switches a and b are connected in series. It is easy to see that this network is in state one iff $A = 1$ and $B = 1$; the network is in state zero iff $A = 0$ or $B = 0$. If we make use of the Boolean algebra $\mathfrak{B}$ composed of the set $\{0, 1\}$ with the usual ordering, the state of this network may be written as

$$A \cap B.$$

Similarly, a necessary and sufficient condition for the network of Fig. 2.2b to be in state one is that $A = 1$ or $B = 1$. In the Boolean algebra $\mathfrak{B}$, the state of this network is

$$A \cup B.$$

We shall confine our attention to networks which can be built up by successive series or parallel connections. There are networks, of great practical importance, which cannot be obtained in this way, but we shall not consider them.

Example 2.1 Which states of the respective switches a, b, c, d, involved in the network of Fig. 2.1, are such that the network is in state one?

Solution: The network of Fig. 2.1 can be obtained by connecting the two networks of Fig. 2.3 in parallel. These two networks can be further broken down into still simpler ones and the results concerning the two simple networks of Fig. 2.2 may be applied to express the states of these networks in terms of the states of the respective switches. The state of the network of Fig. 2.3a is thus found to be

$$(A \cup C) \cap B,$$

while the state of the network of Fig. 2.3b is

$$[(B \cap D) \cup D] \cap B.$$

FIGURE 2.2

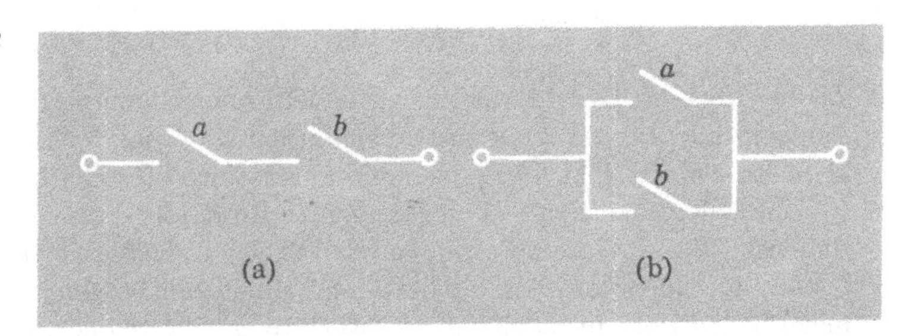

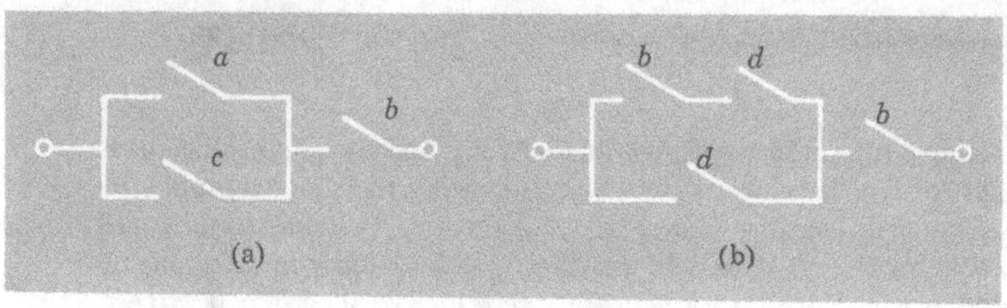

FIGURE 2.3

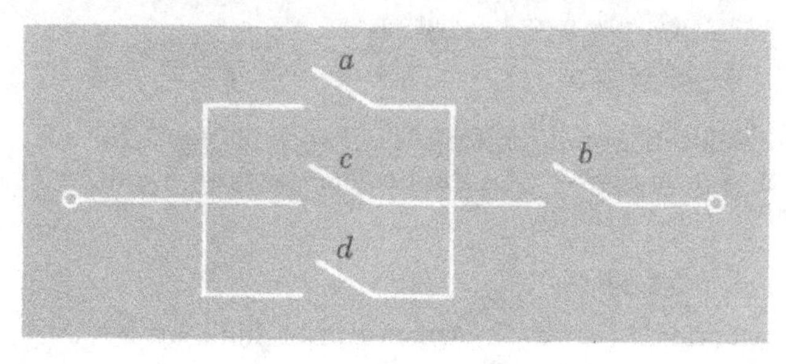

FIGURE 2.4

Since these two networks are to be connected in parallel, the state of the network of Fig. 2.1 is

$$[(A \cup C) \cap B] \cup \{[(B \cap D) \cup D] \cap B\}.$$

A necessary and sufficient condition for the network to be in state one is that the value of this expression is one, in the Boolean algebra $\mathcal{B}$. This work is continued in Example 2.2.

Example 2.2 Find a simpler network which is equivalent to the network of Fig. 2.1.

Solution: From Example 2.1, the state of the network of Fig. 2.1 is

$$[(A \cup C) \cap B] \cup \{[(B \cap D) \cup D] \cap B\}.$$

The expression in the curly bracket simplifies to $B \cap D$. The commutative and distributive laws then give

$$[(A \cup C) \cap B] \cup (B \cap D)$$
$$= [(A \cup C) \cap B] \cup (D \cap B) = (A \cup C \cup D) \cap B.$$

The network of Fig. 2.4 has this state, and is much simpler than the one shown in Fig. 2.1. A necessary and sufficient condition for the network to be in state one is that b be closed and at least one of a, c, d be closed.

Unfortunately, there is no convenient procedure for obtaining the simplest form for an expression in the Boolean algebra $\mathcal{B}$, nor even for testing an expression to see whether or not it can be simplified. Thus, much of the work in replacing a network by a simpler one will depend on individual ingenuity in simplifying the Boolean expression for the state of the network.

Example 2.3 In a committee of three members, each member has control of one of the three switches a, b, c. An issue is to be decided by a simple majority vote, and each member votes on the measure by closing his switch if his vote is "yes" and opening his switch if his vote is "no". Design a network which will be in state one if and only if the measure passes.

Solution: The measure will pass if and only if it receives exactly three votes or exactly two votes; moreover, the two votes for the measure could occur in any one of three different ways. It is now easy to see that the measure will pass iff one of the following four mutually exclusive sets of conditions is satisfied.

$$A = B = C = 1.$$

$$A = B = 1, \; C = 0.$$

$$A = C = 1, \; B = 0.$$

$$B = C = 1, \; A = 0.$$

These four conditions can be rephrased, respectively, as

$$A \cap B \cap C = 1.$$

$$A \cap B \cap \bar{C} = 1.$$

$$A \cap \bar{B} \cap C = 1.$$

$$\bar{A} \cap B \cap C = 1.$$

A necessary and sufficient condition for the measure to pass is that at least one (and, therefore, exactly one) of these conditions be satisfied. Thus, the measure passes iff

$$(A \cap B \cap C) \cup (A \cap B \cap \bar{C}) \cup (A \cap \bar{B} \cap C) \cup (\bar{A} \cap B \cap C) = 1.$$

FIGURE 2.5

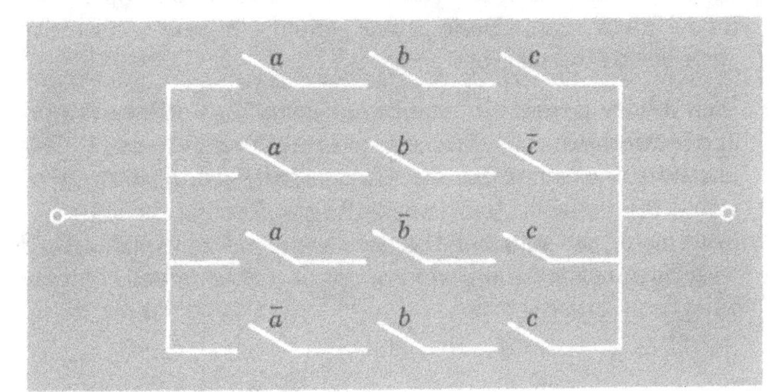

But this means that the Boolean expression

$$(A \cap B \cap C) \cup (A \cap B \cap \bar{C}) \cup (A \cap \bar{B} \cap C) \cup (\bar{A} \cap B \cap C)$$

gives the state of the desired network since the network must be in state one if and only if this Boolean expression has the value one.

A network whose state is given by this Boolean expression is shown in Fig. 2.5; Problem 1 asks for a simpler equivalent network.

PROBLEMS

1. Find a simpler network which is equivalent to the network of Fig. 2.5.

2. For each of the networks shown in Fig. 2.6, find a necessary and sufficient condition that the network be in state one, and find a simpler equivalent network.

3. The game of "matching switches" is played by two people. Each one controls a switch. Design a network which will be in state one iff the switches match (both switches open, or both switches closed).

4. (**a**) In a committee of four men, each man has control over one of the switches a, b, c, d. Each man closes his switch if he votes in favor of a certain measure and opens his switch if he votes against it. Design, and simplify, a network which will be in state one if and only if the measure passes (on a simple majority vote).

   (**b**) Do part (a) if the chairman (controlling switch c) is still given a vote and, in case the vote results in a tie, his vote is used to break the tie.

   (**c**) Do part (a) if the chairman has veto power in addition to his vote and his power to break ties.

5. A small corporation has 100 shares of stock; each share entitles its owner to one vote. The shares are owned by five people in the amounts of 45 shares, 25 shares, 15 shares, 10 shares, and 5 shares respectively. In a vote which is to be decided by a $\frac{2}{3}$ majority, each man has a switch which he closes to vote "yes" for all his shares and opens to vote "no" for all his shares. Design, and simplify, a network which will be in state one iff the measure passes.

6. Would it be possible to change our definitions of the states of switches and networks so that state zero indicates that current *is* allowed to pass and state one means that current is *not* allowed to pass? What changes would this entail in the Boolean function giving the state of a network in terms of the states of the switches in the network? Do Problem 3 with these new meanings for states one and zero, and compare the work with your earlier solution.

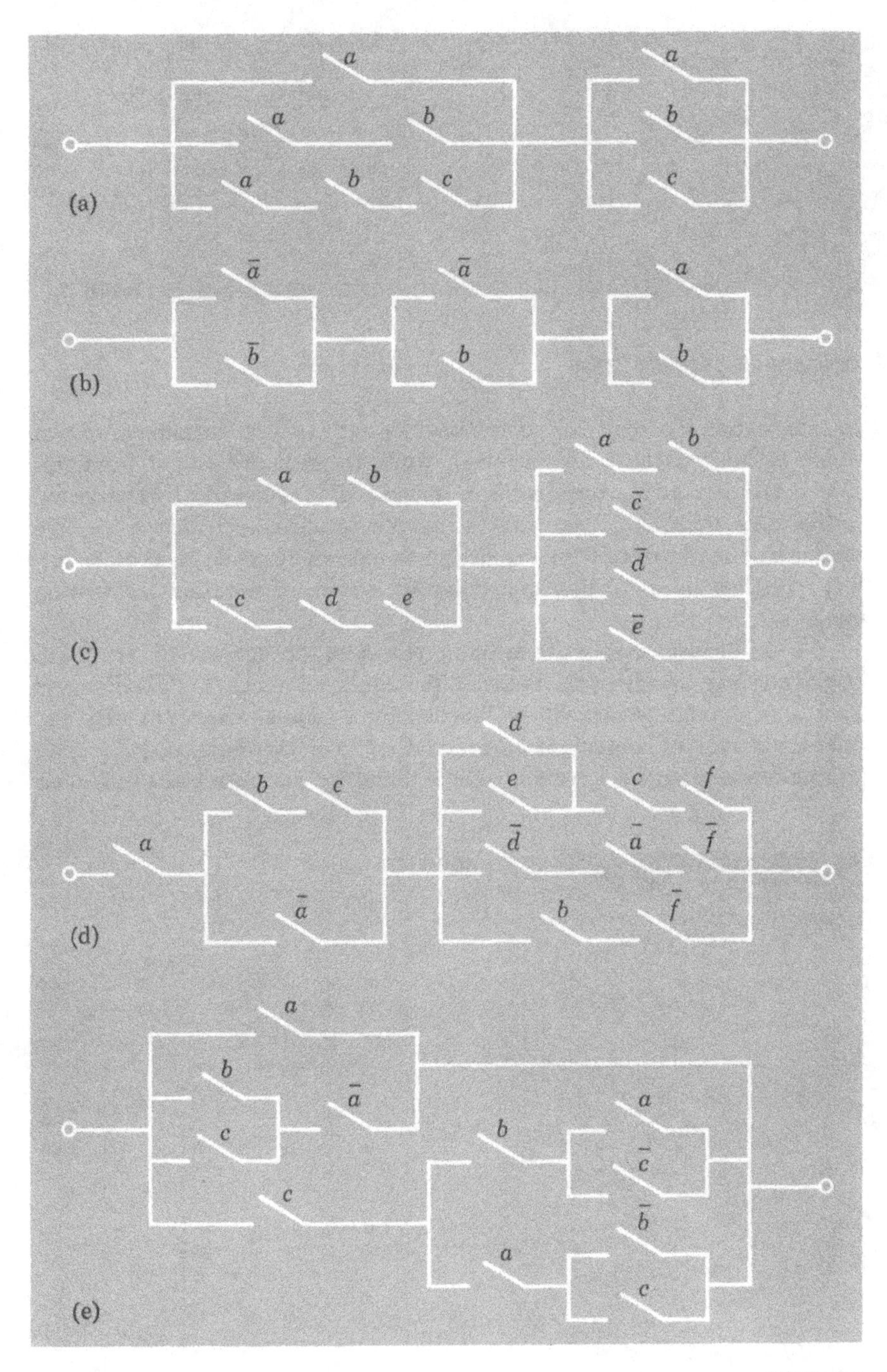
(a)
a
a
b
a
b
c
a
b
c
(b)
$\bar{a}$
$\bar{b}$
$\bar{a}$
b
a
b
(c)
a
b
c
d
e
a
b
$\bar{c}$
$\bar{d}$
$\bar{e}$
(d)
a
b
c
$\bar{a}$
d
e
$\bar{d}$
b
c
f
$\bar{a}$
$\bar{f}$
$\bar{f}$
(e)
a
b
c
$\bar{a}$
c
b
a
$\bar{c}$
$\bar{b}$
a
c

FIGURE 2.6

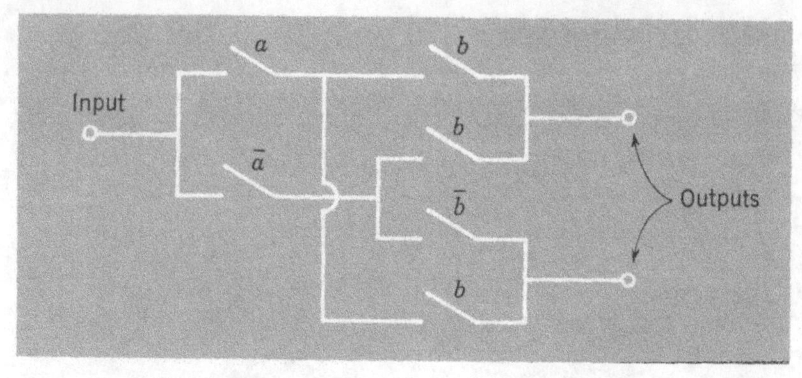

FIGURE 3.1

8-3 Applications to Computer Design

In this section we shall consider networks with several terminals, one of which is labelled the *input* terminal, while the rest are *output* terminals (Fig. 3.1). An output terminal is in *state one* iff it is electrically connected to the input terminal; otherwise, the output terminal is in *state zero*. We shall use networks of this type to design an extremely crude computer, but, first, we shall need to discuss a different system of notation for writing integers.

We shall consider only the integers from zero through seven, but it will be evident how to extend the notation to include all integers. The integers from zero through seven will be labelled by a code as shown in Fig. 3.2. Each code symbol consists of an ordered triple of the digits 0 or 1. The integer represented by a certain code is found by computing the following

Integer	Code
0	000
1	001
2	010
3	011
4	100
5	101
6	110
7	111

FIGURE 3.2

sum (ordinary addition): The number of ones (2^0) given by the right digit, plus the number of twos (2^1) given by the middle digit, plus the number of fours (2^2) given by the left digit. Since the number two is used in this code in the same way as the number ten is used in our ordinary notation for integers [recall that $367 = 3(10^2) + 6(10^1) + 7(10^0)$], this code is called the *binary* notation. As is customary, we shall omit the zeros in the left position, and shall omit them in the middle position for the integers zero and one. Binary notation is frequently used in computer design because only two digits appear in the binary system (the Boolean constants 0 and 1), and many electrical devices have exactly two states.

Addition of numbers in the binary system is easily accomplished by noting that

$$0 + 0 = 0, \quad 0 + 1 = 1 + 0 = 1,$$

$$1 + 1 = 10 \text{ (0 and 1 to "carry").}$$

Three addition problems are shown below. In Example 3.1, we design an electrical network which could be used to add integers in the binary notation.

$$\begin{array}{rrr} 10 & 11 & 1 \\ \underline{101} & \underline{11} & \underline{101} \\ 111 & 110 & 110 \end{array}$$

Example 3.1 Design a network for adding integers using binary notation.

Solution: Suppose we are given two numbers, each written in binary notation, say $A_2\, A_1\, A_0$ and $B_2\, B_1\, B_0$, where A_i and B_i are the binary digits in the two numbers. Let us denote the binary digits in the sum of these two numbers by D_i. Thus, we shall consider the following addition problem (if it is possible).

$$\begin{array}{c} A_2\, A_1\, A_0 \\ \underline{B_2\, B_1\, B_0} \\ D_2\, D_1\, D_0. \end{array}$$

Remember, we are confining our attention to integers between zero and seven, so the sum $D_2\, D_1\, D_0$ must be one of these numbers, otherwise we are not interested in the problem. Of course, the network we design must indicate, in some way, which problems we should reject because the sum is too big.

First, let us design a network which will give us the information we want to obtain from the digits A_0 and B_0. There are two things we want to know.

What is the digit D_0?

Is there a one to "carry"?

To standardize the situation, we shall consider that we always "carry" a digit, but sometimes the digit which is "carried" is zero. Let us use C_0 for the digit which is "carried" from the right position to the middle position. Then we want a network involving switches a_0, b_0, $\bar{a}_0$, $\bar{b}_0$, and with two output terminals, d_0 and c_0, whose states are D_0 and C_0, respectively.

From the binary addition table, we see that D_0 is one iff one of A_0 and B_0 is one and the other is zero. Thus

$$D_0 = (A_0 \cap \bar{B}_0) \cup (\bar{A}_0 \cap B_0).$$

Similarly, C_0 is one iff both A_0 and B_0 are one. Thus

$$C_0 = A_0 \cap B_0.$$

It is now easy to see that the network of Fig. 3.3 gives the information desired from A_0 and B_0.

Now let us turn our attention to the middle digits of the numbers in our addition problem. We shall use C_0, A_1, and B_1 to determine D_1 and C_1 (C_1 is the digit "carried" from the middle to the left position). Thus we want a network involving switches a_1, $\bar{a}_1$, b_1, etc. and having two output terminals, d_1 and c_1, whose states are D_1 and C_1, respectively.

We find (Problem 1)

$$D_1 = \{A_1 \cap [(B_1 \cap C_0) \cup (\bar{B}_1 \cap \bar{C}_0)]\} \cup \{\bar{A}_1 \cap [(B_1 \cap \bar{C}_0) \cup (\bar{B}_1 \cap C_0)]\}$$

and

$$C_1 = [A_1 \cap (B_1 \cup C_0)] \cup (B_1 \cap C_0).$$

The network is shown in Fig. 3.4. In our crude computer, we must observe the value of C_0 from the network of Fig. 3.3, and then manually set the switch c_0 which appears in the network of Fig. 3.4. Of course, in a

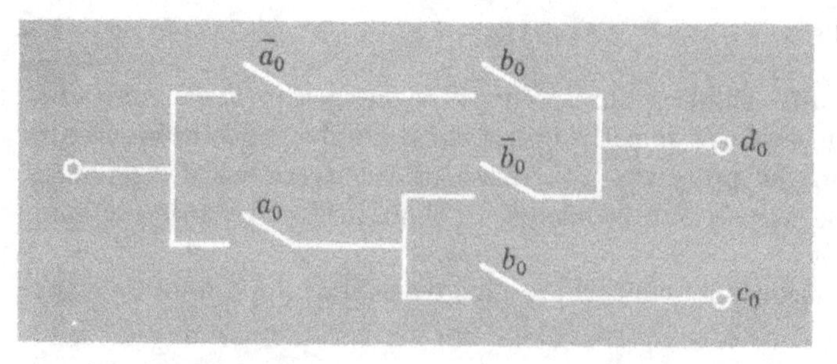

FIGURE 3.3

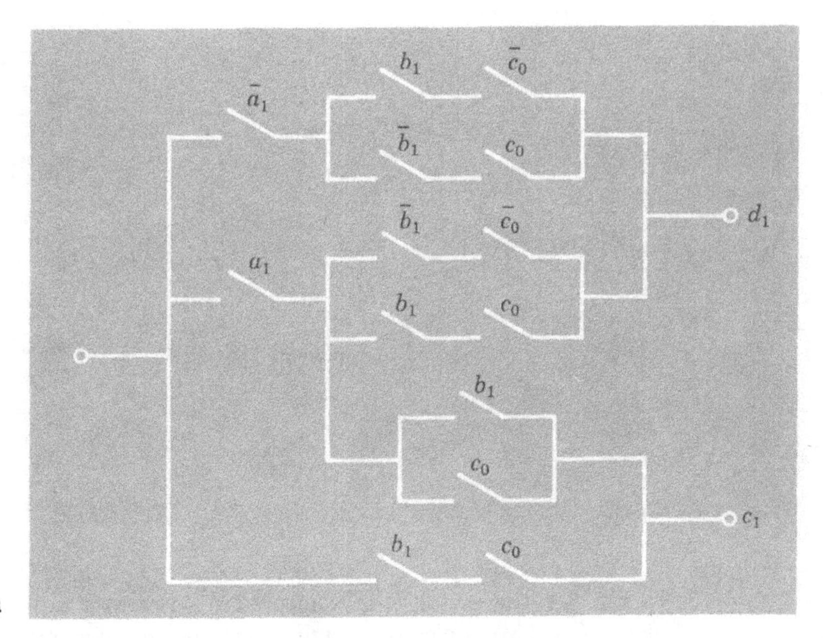

FIGURE 3.4

practical computer, different electrical devices would be used, and the computer itself would do this "carrying."

It is evident that the network of Fig. 3.4, with each subscript increased by one, could be used in combining the left digits A_2 and B_2 with the "carry" C_1 from the middle digits. The states of the two output terminals, D_2 and C_2, would be interpreted as follows. We are interested in the problem iff $C_2 = 0$; in that case, D_2 is the left digit of the sum. Problem 2 suggests a different interpretation of these output states.

PROBLEMS

1. Show that the expressions given for D_1 and C_1 in Example 3.1 are correct.

2. (**a**) Show that, by using an ordered quadruple of Boolean constants instead of the triple discussed in the text, it is possible to write all the integers zero through fifteen in binary notation.

(**b**) Show that the output state C_2, in the computer designed in Example 3.1, may be considered as the leftmost digit in an ordered quadruple thus enabling the computer to add *any* three-digit binary numbers.

In actual operation, a computer usually makes some further use of the numbers which it computes. If these numbers become too big to fit in the registers of the machine, the program for the machine must take account of this fact.

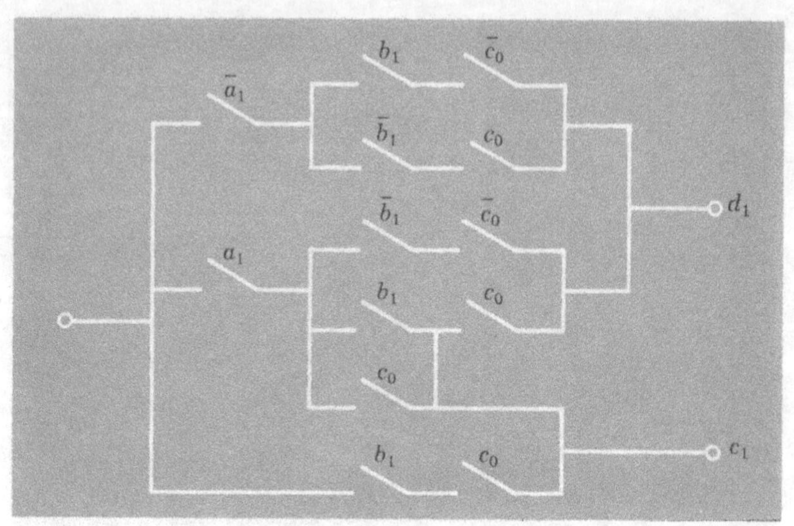

FIGURE 3.5

3. In Fig. 3.4, the switch b_1 is connected in series with a_1 in two different places (i.e. two copies of switch b_1 are used). We might attempt to eliminate this duplication by using only one copy of the switch b_1 and replacing the network of Fig. 3.4 by that shown in Fig. 3.5. Show that the two networks are *not* equivalent.

4. Find Boolean expressions for the states of the two output terminals in the network of Fig. 3.1. Find a simpler equivalent network.

5. Let $A_2\,A_1\,A_0$ and $B_2\,B_1\,B_0$ be two integers between zero and seven, inclusive, written in binary notation. Design a network, involving switches a_0, $\bar{a}_0$, a_1, etc., and having three output terminals g, e, and l. Arrange that g is in state one iff

$$A_2\,A_1\,A_0 > B_2\,B_1\,B_0;$$

e is in state one iff

$$A_2\,A_1\,A_0 = B_2\,B_1\,B_0;$$

l is in state one iff

$$A_2\,A_1\,A_0 < B_2\,B_1\,B_0.$$

6. A student takes a quiz with 3 true-false questions for which the correct answers are "True", "True", "False", respectively. The student has three switches a, b, c, one for each of the questions. For each question, he closes the switch if his answer to the question is "True", and opens the switch if his answer to the question is "False." Design a network with 4 output terminals n_0, n_1, n_2, n_3, such that, for each i $(0 \leq i \leq 3)$, n_i is in state one iff the student answers exactly i of the questions correctly.

7. With reference to Problem 6, design a network with two output terminals, d and e, so that the number of questions the student answers correctly will be given by DE, in binary notation. (D and E are the states of d and e, respectively.)

8-4 Applications to Logic

In this section we shall consider the solution of a system of m equations in n independent variables in a Boolean algebra, and shall consider a method of solving certain problems of the logical-puzzle type. We shall make formal use of the results of Chapter 7.

Let $X_1, X_2, \ldots, X_n$ be n independent variables, each with a given (but arbitrary) Boolean algebra $\mathfrak{B}$ as its range. Let $f_i(X_1, X_2, \ldots, X_n)$ and $g_i(X_1, X_2, \ldots, X_n)$ $(i = 1, 2, \ldots, m)$ be $2m$ formal Boolean expressions formed from these variables and the operations "$\cup$", "$\cap$", and "$^-$".

We consider the set of m equations

$$(*) \qquad f_i(X_1, X_2, \ldots, X_n) = g_i(X_1, X_2, \ldots, X_n). \qquad (i = 1, 2, \ldots, m)$$

A *solution* of this set of equations is an n-tuple $B_1, B_2, \ldots, B_n$ of elements of $\mathfrak{B}$ such that, for each $i(1 \leq i \leq m)$, the two expressions $f_i(B_1, B_2, \ldots, B_n)$ and $g_i(B_1, B_2, \ldots, B_n)$ represent the same element of $\mathfrak{B}$.

Now, let us consider a single equation

$$f(X_1, X_2, \ldots, X_n) = g(X_1, X_2, \ldots, X_n).$$

This equation is equivalent to the two inequalities

$$f(X_1, X_2, \ldots, X_n) \leq g(X_1, X_2, \ldots, X_n)$$

and

$$g(X_1, X_2, \ldots, X_n) \leq f(X_1, X_2, \ldots, X_n);$$

these inequalities, in turn, are equivalent (Problem 6, Section 5-2) to

$$f(X_1, X_2, \ldots, X_n) \cap \overline{g(X_1, X_2, \ldots, X_n)} = L$$

and

$$g(X_1, X_2, \ldots, X_n) \cap \overline{f(X_1, X_2, \ldots, X_n)} = L.$$

Finally, these two equations are equivalent to the single equation

$$[f(X_1, X_2, \ldots, X_n) \cap \overline{g(X_1, X_2, \ldots, X_n)}] \cup [\overline{f(X_1, X_2, \ldots, X_n)} \cap g(X_1, X_2, \ldots, X_n)] = L.$$

Thus, each of the equations in the system (*) can be written with the least element L of $\mathfrak{B}$ as the right member of the equation. But then, evidently, the entire system of equations (*) is equivalent to the single equation

$$\begin{aligned} &[f_1(X_1, X_2, \ldots, X_n) \cap \overline{g_1(X_1, X_2, \ldots, X_n)}] \\ \cup\ &[\overline{f_1(X_1, X_2, \ldots, X_n)} \cap g_1(X_1, X_2, \ldots, X_n)] \cup \ldots \\ \cup\ &[\overline{f_m(X_1, X_2, \ldots, X_n)} \cap g_m(X_1, X_2, \ldots, X_n)] = L. \end{aligned}$$

Now, if we write the expression in the left member of this equation in conjunctive normal form based on $X_1, X_2, \ldots, X_n$ (the conjunction of a set $\mathfrak{S}$ of elementary disjunctions based on $X_1, X_2, \ldots, X_n$), it is easy to see that the equation has a solution iff $\mathfrak{S} \neq \emptyset$ and, when this condition is satisfied, any n-tuple of elements of $\mathfrak{B}$ which makes one of the elementary disjunctions in $\mathfrak{S}$ take on the value L is a solution of the equation. However, this procedure may not give all the solutions, as is shown by Example 4.1.

Example 4.1 Given a Boolean algebra $\mathfrak{B}$, solve the system of equations

$$X_1 = X_2$$

$$X_2 = X_3$$

in $\mathfrak{B}$.

Solution: Using the procedure of the text, we find the following single equation which is equivalent to the given system.

$$(X_1 \cap \bar{X}_2) \cup (\bar{X}_1 \cap X_2) \cup (X_2 \cap \bar{X}_3) \cup (\bar{X}_2 \cap X_3) = L.$$

Writing the left member in conjunctive normal form based on X_1, X_2, X_3, gives

$$(X_1 \cup X_2 \cup X_3) \cap (\bar{X}_1 \cup \bar{X}_2 \cup \bar{X}_3) = L.$$

Thus, the system of equations has solutions. One of its solutions is given by

$$X_1 = X_2 = X_3 = L,$$

and another solution is given by

$$\bar{X}_1 = \bar{X}_2 = \bar{X}_3 = L$$

or, equivalently,

$$X_1 = X_2 = X_3 = G.$$

Of course, there are as many more solutions as there are additional elements of $\mathfrak{B}$.

We turn our attention now to certain problems in logic. We suppose the statements $a_1, a_2, \ldots, a_n$ are given, but that we may not know the respective truth values $A_1, A_2, \ldots, A_n$ of these statements. However, we are given a set of Boolean equations (in the Boolean algebra $\{0, 1\}$) involving $A_1, A_2, \ldots, A_n$. The problem is to find these truth values.

The procedure discussed above gives a method for solving such problems, but the algebraic manipulations may become cumbersome. They can frequently be simplified with a little ingenuity. Illustrative examples are presented in Examples 4.2 and 4.3. For a more detailed discussion, see Refs. 10, 4, and 19. The last two of these references discuss the question of uniqueness of the solutions.

Example 4.2 Out of six students, exactly two are known to have been cheating. Andy says "Ben and Charles". David says "Edgar and Fred". Edgar says "Ben and Fred". Charles says "Andy and Ben". Ben says "David and Edgar".

In one of these quotations, neither boy mentioned has been cheating. In each of the other four quotations, one of the boys mentioned has been cheating and the other has not. Who are the cheaters?

Solution: We shall use a for the statement "Andy has been cheating", and similarly for b, c, d, and e. The truth value of each of these statements will be denoted by the corresponding capital letter, and we shall find equations in the Boolean algebra $\{0, 1\}$.

Each quotation is a lie, since at least one of the boys named has not been cheating. Thus

$$B \cap C = E \cap F = B \cap F = A \cap B = D \cap E = 0. \tag{1}$$

Also, in one of the quotations, neither boy mentioned has been cheating. Hence

$$(B \cup C) \cap (E \cup F) \cap (B \cup F) \cap (A \cup B) \cap (D \cup E) = 0. \tag{2}$$

Finally, for some set of four of the quotations, each one of the four gives one of the cheaters correctly. Thus

$$\begin{aligned} &[(E \cup F) \cap (B \cup F) \cap (A \cup B) \cap (D \cup E)] \\ \cup\ &[(B \cup C) \cap (B \cup F) \cap (A \cup B) \cap (D \cup E)] \\ \cup\ &[(B \cup C) \cap (E \cup F) \cap (A \cup B) \cap (D \cup E)] \\ \cup\ &[(B \cup C) \cap (E \cup F) \cap (B \cup F) \cap (D \cup E)] \\ \cup\ &[(B \cup C) \cap (E \cup F) \cap (B \cup F) \cap (A \cup B)] = 1. \end{aligned} \tag{3}$$

Using the distributive law and Eqs. (1), we find that Eq. (2) reduces to

$$(B \cap E) \cup (C \cap F \cap A \cap D) = 0.$$

Since this disjunction is zero, each of the terms is zero.

$$B \cap E = 0. \tag{4}$$

(We already know that $C \cap F \cap A \cap D = 0$, since only two of the boys have been cheating; this remark will be used in simplifying Eq. (3) below.)

Now using the distributive law and Eqs. (1) and (4) to simplify Eq. (3), we obtain

$$B \cap D = 1.$$

Ben and David have been cheating.

Example 4.3 Ann, Betty, Cora, and Doris have engaged in a cake-baking contest and each won one of the four prizes. Three of the girls reported as follows. Ann said "Cora was first; Betty was second". Betty said "Cora was second; Doris was third". Cora said "Ann was second; Doris was fourth". Each of the three girls made one correct statement and one false one. What were the results of the contest?

Solution: We use a_1 for the statement "Ann was first", and similarly for b_2, c_4, etc.; again, the corresponding capital letters are the truth values of these sentences.

Since one of Ann's statements is true, and the other is false,

$$(C_1 \cap \bar{B}_2) \cup (\bar{C}_1 \cap B_2) = 1.$$

Treating the other two quotations similarly, and taking a conjunction,

$$\begin{aligned} &[(C_1 \cap \bar{B}_2) \cup (\bar{C}_1 \cap B_2)] \cap [(C_2 \cap \bar{D}_3) \cup (\bar{C}_2 \cap D_3)] \\ &\cap [(A_2 \cap \bar{D}_4) \cup (\bar{A}_2 \cap D_4)] = 1. \end{aligned} \tag{1}$$

From the nature of the problem, no girl won two different prizes, and no prize was won by two different girls. Thus, we have the conditions

$$C_1 \cap C_2 = 0, \quad B_2 \cap C_2 = 0, \text{ etc.} \tag{2}$$

Using the distributive law, and Eq. (2), we find, from Eq. (1),

$$(C_1 \cap \bar{B}_2) \cap (\bar{C}_2 \cap D_3) \cap (A_2 \cap \bar{D}_4) = 1.$$

From this conjunction we find that Cora was first, Ann was second, and Doris was third. Therefore, Betty was fourth.

PROBLEMS

1. Show that any system of equations in a Boolean algebra $\mathcal{B}$ is equivalent to a single equation in which the right member is the greatest element G.

2. Show that if an equation in a Boolean algebra has any solutions, then it has a solution in which each variable has one of the values L or G.

3. For each of the following systems of equations in a Boolean algebra $\mathcal{B}$, find a solution or show that there is none.

(a)
$$X_1 = X_2 \cap X_3$$
$$X_2 = X_1 \cup \bar{X}_3$$
$$X_3 = \bar{X}_2 \cap \bar{X}_1.$$

(b)
$$X_1 = X_2 \cap X_3$$
$$X_2 = X_1 \cup X_3$$
$$X_3 = \bar{X}_2 \cap \bar{X}_1.$$

4. One of the four men Adams, Baker, Carlson, and Davis has committed a crime. The men make the following statements.

Adams: "Carlson did it "

Baker: "I didn't do it."

Carlson: "Adams' statement is false."

Davis: "Adams did it."

(a) If exactly one of the four statements is true, who is the guilty man?

(b) If exactly one of the four statements is false, who is guilty?

5. Three marbles are drawn successively and at random from a sack containing marbles which are red or white or blue. Two of those drawn are red and exactly one of the following three statements is true.

(1) The first marble drawn is blue.

(2) The second marble drawn is red.

(3) The third marble drawn is not white.

What is the color of each of the three marbles drawn?

6. Four blocks of wood are sitting on a table. The top face of each block is colored red, green, yellow, or blue, and no two are colored the same. The bottom faces of the blocks are also colored with the same four colors; no two of the bottom faces are the same color, and no block has the same color on both its top and bottom faces. The block with the red top face does not have a blue bottom face. For the block with the yellow top face, the color of its bottom face is the same as the color of the top face of the block for which the bottom face has the same color as the top face of the block whose bottom face is green. What are the colors of the four blocks?

REFERENCES

Chapter 1: 1, 3, 6, 7, 8, 9, 11, 12, 14, 16, 17, 18, 20, 23, 24, 25, 26, 28, 29, 31.

Chapter 2: 6, 7, 8, 25.

Chapter 3: 2, 5, 7, 25, 26.

Chapter 4: 5, 25.

Chapter 5: 1, 2, 5, 7, 25, 26, 27, 30, 31.

Chapter 6: 7, 25, 27, 31.

Chapter 7: 14, 21, 22, 25, 30, 31.

Chapter 8: 1, 2, 3, 4, 8, 10, 12, 13, 19, 31, 32.

1. Allendoerfer, C. B. and C. O. Oakley, *Principles of Mathematics.* New York: McGraw-Hill Book Company, Inc., 1955. (Freshman text.)

2. Andree, R. V., *Selections from Modern Abstract Algebra.* New York: Henry Holt and Company, 1958. (Chapter 3 gives a simple, readable, elementary discussion of Boolean algebra.)

3. Berkeley, Edmund C., *Symbolic Logic and Intelligent Machines.* New York: Reinhold Publishing Corporation, 1959. (Applications of Boolean algebra to the design of electrical networks.)

4. Bernstein, B.A., "Note on the Condition that a Boolean Equation have a Unique Solution," *American Journal of Mathematics*, vol. **54** (1932) pp. 417–18. (Simple proofs of two theorems of Whitehead (Ref. 30) on solution of an equation in a Boolean algebra.)

5. Birkhoff, Garrett and S. Maclane, *A Survey of Modern Algebra.* New York: The Macmillan Company, 1953. (A senior, or beginning graduate, text. Chapter 11 is concerned with Boolean algebras, ordered sets, and lattices.)

6. Boole, George, *An Investigation of the Laws of Thought.* New York: Dover Publications, Inc., 1951. (A reprint of the classical work by the originator of Boolean algebra; of historical interest.)

7. Carroll, Lewis, *Symbolic Logic and the Game of Logic.* New York: Dover Publications, Inc., 1958. Paper back. (An excellent, elementary presentation of the rudiments of logic, with applications of the recreational type.)

8. Christian, R. R., *An Introduction to Logic and Sets.* New York: Ginn and Company, 1958. Paper back. (An elementary introduction to basic concepts. Preliminary edition.)

9. Exner, R. M. and M. J. Rosskopf, *Logic in Elementary Mathematics.* New York: McGraw-Hill Book Company, Inc., 1959. (The logical structure of selected portions of elementary mathematics is discussed. However, the first portion of the book, on logic, must be understood before these discussions can be read with profit.)

10. Fletcher, T. J., "The Solution of Inferential Problems by Boole Algebra," *The Mathematical Gazette*, vol. **35** (1952) pp. 183–88. (Solution of problems of the logical puzzle type.)

11. Henkin, Leon, W. Norman Smith, Verne J. Varineau, and Michael J. Walsh, *Retracing Elementary Mathematics.* New York: The Macmillan Company, 1962. (Includes a lucid discussion of mathematical induction and its axiomatic foundation.)

12. Hohn, Franz E., *Applied Boolean Algebra: An Elementary Introduction.* The Macmillan Company, 1960. Paper back. (A preliminary edition of the first portion of a forthcoming book. Applications of Boolean algebra to switching circuits and computers are emphasized. The book gives a simple discussion of circuits, logic, and subsets of a set as examples of Boolean algebras. There is some discussion of engineering and hardware problems.)

13. ————, "Some Mathematical Aspects of Switching," *The American Mathematical Monthly*, vol. **62** (1955) pp. 75–90. (Includes an extensive bibliography of works dealing with applications of Boolean algebra.)

14. Hilbert, D. and W. Ackerman, *Principles of Mathematical Logic.* New York: Chelsea Publishing Company, 1950. (A quite readable text on elementary logic.)

15. Jacobson, N., *Lectures on Abstract Algebra*, vol. I. Princeton, N. J.: D. Van Nostrand Company, Inc., 1951. (A beginning graduate text. Chapter 7 discusses ordered sets, lattices, Boolean algebras and rings.)

16. Kemeny, John G., Hazleton Mirkil, J. Laurie Snell, and Gerald L. Thompson, *Finite Mathematical Structures.* Englewood Cliffs, N. J.: Prentice-Hall, Inc., 1959. (Chapter 1 is similar to reference 17 below.)

17. Kemeny, John G., J. Laurie Snell, and Gerald L. Thompson, *Finite Mathematics.* Englewood Cliffs, N. J.: Prentice-Hall, Inc., 1956. (Chapter 1 presents an excellent discussion of statements, implication, methods of proof, etc.)

18. Landau, E., *Foundations of Analysis*, trans. F. Steinhardt. New York: Chelsea Publishing Company, 1957. (This book presents, in detail, the construction of the real and complex number systems starting from Peano's postulates. Mathematical induction is used extensively.)

19. Parker, W. L. and B. A. Bernstein, "On Uniquely Solvable Boolean Equations," *University of California Publications in Mathematics*, vol. **3** (1955) pp. 1–30. (Several necessary and sufficient conditions are given for a Boolean equation to have a unique solution.)

20. Polya, G. *How to Solve It.* Garden City, N. Y.: Doubleday & Company, Inc., 1957. Paper back. (An excellent discussion of methods of proof, written by a master expositor.)

21. Quine, W. V., "A Way to Simplify Truth Functions," *The American Mathematical Monthly*, vol. **62** (1955) pp. 627–30. (Continuation of Ref. 22.)

22. ————, "The Problem of Simplifying Truth Functions," *The*

American Mathematical Monthly, vol. **59** (1952) pp. 521–30. (References 21 and 22 describe Quine's standard form for an expression in a Boolean algebra, and give a procedure for writing an element in that form.)

23. Richardson, Moses, *Fundamentals of Mathematics*. New York: The Macmillan Company, 1958. (Freshman text.)

24. Rosser, J. Barkley, *Logic for Mathematicians*. New York: McGraw-Hill Book Company, Inc., 1953. (Symbolic logic.)

25. Stabler, E. R., *Introduction to Mathematical Thought*. Reading, Massachusetts: Addison-Wesley Publishing Company, Inc., 1953. (Presents almost all of our material, but frequently from a different viewpoint, and in a different order.)

26. Stoll, R. R., *Sets, Logic and Axiomatic Theories*. San Francisco: W. H. Freeman and Company, 1961. (Paper back. A good, readable presentation; includes material on relations and on Boolean algebra.)

27. Stone, M. H., "The Theory of Representations for Boolean Algebras," *American Mathematical Society Transactions*, vol. **40** (1936) pp. 37–111. (A classic paper. One of Stone's results is that every Boolean algebra (finite or not) is isomorphic to a Boolean algebra of sets.)

28. Suppes, Patrick, *Introduction to Logic*. Princeton, N. J.: D. Van Nostrand Company, Inc., 1957. (Symbolic logic.)

29. Tarski, A., *Introduction to Logic*. New York: Oxford University Press, 1939. (Simpler and more readable than refs. 9, 24, 28.)

30. Whitehead, A. N., "Memoir on the Algebra of Symbolic Logic," *American Journal of Mathematics*, vol. **23** (1901) pp. 139–65, 297–316. (A classic paper on the structure of a Boolean algebra. Solution of equations is discussed, but see also Refs. 4 and 19.)

31. Whitesitt, J. Eldon, *Boolean Algebra and its Applications*. Reading, Massachusetts: Addison-Wesley Publishing Company, Inc., 1961. (A very good treatment of Boolean algebra including elementary set theory and logic; considerable discussion of applications to electrical networks.)

32. Young, F. H., *Digital Computers and Related Mathematics*. New York: Ginn and Company, 1961. Paper back. Applications of Boolean algebra to the design and programming of computers.

INDEX

CPSIA information can be obtained
at www.ICGtesting.com
Printed in the USA
LVHW04s2157140918
590038LV00001B/121/P